HUMAN ERROR:
Cause, Prediction, and Reduction

SERIES IN APPLIED PSYCHOLOGY

Edwin A Fleishman, George Mason University
Series Editor

Psychology in Organizations: Integrating Science and Practice
Kevin R. Murphy and Frank E. Saal

Teamwork and the Bottom Line: Groups Make a Difference
Ned Rosen

Patterns of Life History: The Ecology of Human Individuality
Michael D. Mumford, Garnett Stokes, and William A. Owens

Work Motivation
Uwe E. Kleinbeck, H.-Henning Quast, Henk Thierry, and Hartmut Häcker

Human Error: Cause, Prediction, and Reduction
John W. Senders and Neville P. Moray

HUMAN ERROR:
Cause, Prediction, and Reduction

analysis and synthesis by
JOHN W. SENDERS, PH.D.
NEVILLE P. MORAY, D.PHIL.

CRC Press
Taylor & Francis Group
Boca Raton   London   New York

CRC Press is an imprint of the
Taylor & Francis Group, an **informa** business

First Published 1991 by
Lawrence Erlbaum Associates, Inc.

Published 2019 by CRC Press
Taylor & Francis Group
6000 Broken Sound Parkway NW,
Suite 300 Boca Raton, FL 33487-2742

© 1991 by Taylor & Francis Group, LLC
CRC Press is an imprint of Taylor & Francis Group, an Informa business

First issued in paperback 2019

No claim to original U.S. Government works

ISBN-13: 978-0-367-45064-9 (pbk)
ISBN-13: 978-0-89859-598-7 (hbk)

Library of Congress Cataloging-in-Publication Data

Human error (cause, prediction, and reduction) : analysis and sythesis / by John W. Senders, Neville P. Moray.
    p. cm. -- (Series in applied psychology)
    Chiefly papers presented at the Second Conference on the Nature and Source of Human Error, sponsored by NATO, held 1983, at Bellagio Italy.
    Includes bibliographical references and index.
    ISBN 0-89859-598-3
    1. Errors--Congresses. 2. Error--Congresses. I. Senders, John W., 1920- . II. Moray. Neville. III. Conference on the Nature and Source of Human Error (2nd : 1983 : Bellagio, Italy) IV. North Atlantic Treaty Organization. V. Series.
BF323.E7H86 1991
153.7'4--dc20
                                90-27312
                                    CIP

Contents

Foreword

There is a compelling need for innovative approaches to the solution of many pressing problems involving human relationships in today's society. Such approaches are more likely to be successful when they are based on sound research and applications. This Series in Applied Psychology offers publications which emphasize state-of-the-art research and its application to important issues of human behavior in a variety of societal settings. The objective is to bridge both academic and applied interests.

Few would dispute the fact that biographical data predict performance and the importance of the need better to understand the nature of human error. It has been variously estimated that as many as 90 percent of industrial and system failures are triggered by human error. Yet relatively little attention has been paid to error by behavioral scientists. The need for such attention struck John Senders in 1978 as a serious flaw in the programs of research on human performance supported by the Department of Defense. He urged those doing such research to pay attention to error, and he took the advice himself. Neville Moray needed little convincing, and the effort to improve knowledge of human failure has become somewhat of a crusade for both.

They have worked and taught over the last 20 years predominantly in the context of engineering and, naturally enough in that context, tended to emphasize the practicality and utility of the models and theories of behavioral science. The study of human performance has had a compelling and continuous attraction for both; and both are deeply committed to the reduction of the risk and the cost associated with human error. In the last 10 years they have increasingly devoted their efforts to an understanding of error as a psychological topic in itself rather than as an index or measure of the effect of changes in the work place or in methods of training.

Earlier, Senders and Moray were instrumental in bringing together individuals from a variety of behavioral science disciplines who were concerned with issues in conceptualizing human error, in carrying out research on factors contributing to such errors, and in reducing errors in operational settings. These individuals participated in a series of conferences held in Columbia Falls, Maine, and in Bellagio, Italy, in the early 1980s. For the purpose of this book, all participants were asked to update their views and the editors integrated these responses with their own current observations and views on the issues.

Their thinking, and that of the persons who participated in these meetings, is presented in this volume as a stimulus—a stimulus to study human error, to understand something of why it occurs, and to understand something of what can and what cannot be done about it.

Edwin A. Fleishman, Editor
Series in Applied Psychology

Preface

This book is the result of a detailed analysis and a subsequent synthesis of the contributions of a large number of individuals who have been thinking about human errors for a long time. These individuals participated in a series of conferences.

The first conference, called The Clambake Conference on the Nature and Source of Human Error, was held in 1980 in Columbia Falls, Maine. The meeting was conceived and organized by John W. Senders and Ann Crichton-Harris. Despite the recency of the events of Three Mile Island, there was no support to be found from any external agency. The participants paid their own way. Local residents of Columbia Falls opened their homes and provided inexpensive bed-and-breakfast services. The Town of Columbia Falls generously gave us the use of the High School building, long without students and, at the time, devoted primarily to Bingo games. The losses were absorbed by Senders and Crichton-Harris.

The second conference, called The Conference on the Nature and Source of Human Error of 1983, from which most of the material in this book came, was proposed by Senders and Moray to the North Atlantic

Treaty Organization (NATO). NATO sponsored and
supported the meeting with enthusiasm. They provided
money for the travel of the participants and for ad-
ministrative expenses. The meeting was further supported by the Rockefeller Foundation, who provided
the facilities of the Villa Serbelloni, in Bellagio, Italy,
for the meeting to be held. Once we arrived at
Serbelloni, everything was provided. We, and the
other participants, are most grateful to NATO and the
Rockefeller Foundation, and to the staff of the Villa
Serbelloni, for their support.

Each participant submitted a position paper in response to our list of queries prior to these meetings and
participated in the preparation of a group position
paper on each of a group of selected topics after the
meetings. As analysts and synthesists, we have dissected
what was offered, extracted and identified the common
elements, and reassembled them into what we hope is
a coherent structure. The ideas presented here are
those of all the participants. We have acted as the
agents.

This book has been a long time in the making. That
is our fault. We can only plead the usual kinds of
academic movings about which interfere with productive work. The time has not all been wasted. It was
inevitable that, in the process of digesting and filtering
the ideas of the participants (even our own ideas),
concepts which have arisen as a consequence of the
conference and the exchanges that took place there
would find their way into our thoughts. This has some
benefit, of course; the finished product is very nearly
up to date. Progress in the study of error is slow—the
topic is not a simple one.

In an effort to mitigate the lapse of time, we invited all the participants to write their current thoughts on the topic to be included in the text. For most, it appeared that there were no changes to be recorded. A few of the participants, however, did respond at length to our request for new thoughts on human error, and we have included their contributions as appendices. A few others made lesser comments, and we have incorporated them into the main body of the text.

We acknowledge gratefully the work of Ann Crichton-Harris, who provided organization for the Bellagio conference, as she had for the conference at Columbia Falls.

List of Contributors*

Dr. David Embry
Dr. Klaus-Martin Goeters
Dr. Martine Griffon-Fouco
Dr. Erik Hollnagel
Prof. Ezra S. Krendel
Prof. Elizabeth F. Loftus
Dr. Giovanni Mancini
Mr. Duane McRuer
Prof. Neville P. Moray
Prof. Donald A. Norman
Prof. Orlindo Pereira

Prof. Jens Rasmussen
Prof. James Reason
Dr. William B. Rouse
Prof. William Ruddick
Prof. John W. Senders
Prof. Thomas B. Sheridan
Dr. Alan D. Swain
Dr. David Taylor
Prof. Willem A. Wagenaar
Prof. David Woods
Dr. John Wreathall

---

* Institutional affiliations and addresses are listed for each contributor in Appendix II

Introduction 1

THE STUDY OF HUMAN ERROR

All of us have experienced human error. When we
interact with machines or complex systems, we frequently do things that are contrary to our intentions.
Depending on the complexity of the system and the
intentions of the people interacting with it, this can be
anything from an inconvenience (often it is not even
noticed) to a genuine catastrophe. Human error can
occur in the design, operation, management, and
maintenance of the complex systems characteristic of
modern life. Because we depend increasingly on these
systems for our well being, it is clear that human error
is a potent and frequent source of hazard to human life
and welfare, and to the ecosystems of Earth.

It is only necessary to read the newspapers to see
that human error is of crucial importance; it is a crucial
factor in many of the great and devastating events
which capture our attention—Bhopals, Chernobyls,
Three Mile Island. But errors that lead to spectacular
accidents of enormous size and cost are merely the tip
of the iceberg. Millions of other errors are made every
day around the globe. Fortunately, they do not all

1

result in disaster. But this is not so much the consequent of good planning as of mere good luck. Those
who have studied the impact of error on complex
human-machine systems generally believe that between
30% and 80% of serious incidents are due, in some
way, to human error.

In September of 1983, twenty-two scientists of various disciplines met for a week to discuss human error.
They came from seven countries to a conference in
Bellagio, Italy sponsored by the Science Committee of
NATO, with assistance from the Rockefeller Foundation. Among the participants were psychologists,
philosophers, mathematicians, and engineers. This
diverse group drew on a wide variety of traditions in
studying human nature, and the interaction of human
beings with the technological systems they use. This
book is the result. We hope it will be read and understood by many people in government, industry, and
society at large. These individuals and the organizations
of which they are members must incorporate in their
decision-making processes consideration of why errors
occur, what errors are likely to occur, and what may be
done about them.

This is difficult. Thus far, there has been very little
systematic study of human error. One reason for this
is that error is frequently considered only as a *result* or
a *measure* of some other variable, and not as a phenomenon in its own right. As Bhopal, Three Mile
Island, Chernobyl, and the many aerospace accidents
make clear, the need for such study is urgent. We think
that our decision-makers should learn what is known
about the nature and source of human error; we also

hope that this introduction to the present state of our knowledge will promote a climate in which they will see the need to finance and undertake further and more extensive research.

HOW THIS BOOK CAME TO BE

In 1978 the United States Air Force (USAF) Office of Scientific Research sponsored a review of research on man-machine systems, which this book's first author attended. He listened for two days to reports of excellent research projects and progress on them. At that point, in a sudden, surprising reformulation of perception, he recognized that virtually all the researchers were studying and analyzing only *correct* behavior, whereas it was self-evident that most of the problems faced by the military in general (and by the USAF in particular) stemmed from the *incorrect* behavior of both military and nonmilitary personnel. The sudden awareness of what appeared to be quite a serious deficiency in man-machine systems research led him to search the literature for information relating to the underlying mechanisms of human error.

He was able to identify relatively little systematic research or theory. The work of Freud (1914) was of course familiar, but it seemed unreasonable to search for the roots of pilot error in the subconscious, the unconscious, or the psycho-sexual state of the pilot. It appeared more useful and reasonable to attribute errors either to aspects of the pilot's environment or to some causal mechanism in the pilot.

But, as everybody seemed to say, "to err is human." If by this they meant that errors will happen from time to time and there is nothing anyone can do about it, what was there for him to study? *Could there be a science of error and error reduction?* Even looking back several decades in the history of modern psychology he found a remarkable lack of interest in the study of error. There are a few interesting suggestions here and there, however.

One of the earliest studies was that of Kollarits (1937) who examined approximately 1200 errors committed by himself, his wife, and his colleagues. He proposed a fourfold classification based on the superficial appearance of the error (what is called a "phenomenological" classification or taxonomy). The categories he used were: *substitution*, *omission*, *repetition*, and *insertion*. He observed that it was sometimes difficult to assign an error unequivocally to a single category, as all who subsequently classified errors in this way have discovered. The point where an error of substitution ends and one of insertion begins, for example, is not always clear. While Kollarits' work was a useful starting place, it did not seriously consider the mechanisms and processes which underly errors—only the form of the errors themselves. In his study, it is interesting to note that the frequencies with which various types of error occurred seem to remain consistent over quite long periods. This suggests that there may indeed be stable and systematic causes of errors, which could be discovered. (Similar results have been found in several more modern studies.)

There was little of theoretical interest beyond this. Spearman (1928) said in 1927: "It is as if psychologists deliberately avoided the question of error." As far as could be determined from a search of the literature, not much had changed in the intervening half century.

SOME CRUCIAL QUESTIONS
ABOUT HUMAN ERROR

Do errors have causes? This question is fundamental to a research program on human error. If there are no causal mechanisms, then we can do very little except to study the statistics (as researchers have done in the past), and examine the circumstances under which error rates vary. This might reveal consistent relationships between the frequency of errors and particular environmental, psychological, and physiological circumstances.

We are all aware of such consistent relationships. Early in January most of us find that we use the old instead of the new year on checks, letters, and other documents, and in our speech as well. In the course of time this tendency fades away, to be renewed the following January. The existence of these rate changes makes it highly likely that at least *some* errors are caused.

There is another important aspect to the question of cause. How many possible causes are there? If every error has its own unique cause, the practical designer of complex systems faces insuperable problems. Each conceivable error would require its own analysis; a remedy for one error would not apply to any other. On the other hand, if there are relatively few causal mechanisms, we can apply some general rules repeatedly, to good effect.

Some errors may be caused by activity in the central nervous system (CNS) that can be isolated and identified. If so, there is a possibility of detecting the CNS activity and responding to it in time to prevent the error, or to reduce its consequences. Needless to say, this would open up many interesting areas of research. Even if errors are random, they may still be caused by random events in the CNS. Then, given increasingly sophisticated technology to measure such neural events, the possibility arises of eliminating errors, or at least taking steps to provide assistance at critical moments.

However, if errors are random events with *no* cause, then it will *always* be impossible to predict either the exact moment (or even the approximate time) when any single error will occur or the form it will take. Only rates will be predictable. Although empirical error rates can be estimated from the few data available, there is no way to be sure that they will stay constant from moment to moment, or from day to day.

Even if we do not know when an error will occur, can we predict what form it may take? This question obviously has tremendous implications for system design. It might be possible to erect defenses that would prevent at least some errors from happening. Such defenses could create a trade-off in which other, less undesirable errors become more likely. Alternatively, machines and systems could be designed to "absorb" the errors that *would* be made. As we shall see, there is some skepticism among the experts in the field as to whether the forms of errors could be predicted, since so few data are available.

Many millions of errors have been noticed and recorded in everyday tasks as well as in psychological research. Generally, however, it has been the custom in research to use *correct* performances as an index of a system's quality or an actor's proficiency, and to ignore *incorrect* performances except for the mere counting of them. Thus, despite the accumulation of data from thousands of experiments, we know only that something was done wrong on many occasions, and virtually nothing about *what* was done, *how* it was done, and *when* it was done. The lack of such data greatly impedes the formulation of any general theory of human error.

THE DISCIPLINE DEVELOPS: THE COLUMBIA FALLS CONFERENCE

The accident at Three Mile Island in 1979 focused the attention of the media and the public on human error in a dramatic and memorable way. In July 1980, an informal conference was held in Columbia Falls, Maine. As far as we know, it was the first to concern itself entirely with fundamental questions relating to human error. Eighteen scientists representing many disciplines attended, coming from the United States, Canada, and England. A list of the attendees appears as Appendix 1. Since several of the participants did not wish to commit themselves to ideas expressed in such an informal forum, no proceedings were published .

All those who were invited to the original conference in Maine had been asked to provide written answers in advance to several questions, including the following:

1. What is an error?
2. Are errors caused?
3. If so, are there 1, 2, 3 or an infinity of causes?
4. Is there a recognizable state of the central nervous system (CNS) of the actor[1] prior to the emission of an error?
5. Do errors occur randomly? Or can the time of an error be predicted?
6. Can the form of an error be predicted?
7. Would it be desirable to eliminate all human error or is error related to creativity?

There were good reasons for all of these queries: (1) It was necessary for people to start talking about what they meant by the words they used. (2) The problem of causality had to be faced rather than being swept aside by a flood of experimental data. If errors could be said to be uncaused, the legal and ethical implications would be enormous. (3) If each error had its own cause, the problems for the researcher as well as for the decision maker would be amplified. (4) If there were a recognizable antecedent state of the CNS, there might be a way of detecting, and absorbing or neutralizing, errors before they occurred. (5) Even if errors are caused, the causes might be random, and predicting exactly when they would occur would be impossible. (6) Even if it is impossible to predict the

---

[1] Throughout this book we will use the word "actor" to mean "one who performs an action of any kind," and not a stage artist. We find this terminology useful, as much of the discussion centers on correct or incorrect "acts" or "actions" performed by such "actors."

time of an error, it might be possible to predict the form that an error will take. This follows from the fact that actions are drawn from the available repertoire of things to be done. (It is impossible to push a button that is not there.) (7) If creativity and error are opposite sides of the same coin (unplanned variation in performance), then eliminating error, if that were possible, might also inhibit creative problem-solving.

The conference in Columbia Falls was productive in many ways, achieving many of the goals set by the organizers. Professional links were established between many people working in the field who had not known one another; in several cases this led to cooperative research in the ensuing years. In general the participants were more than ever convinced of the necessity for scientific study of error as a behavioral phenomenon in its own right rather than simply as an index of performance. Among the participants, however, there was considerable reluctance to discuss theory as to the underlying nature of error-producing mechanisms. Even more strongly felt was a reluctance to discuss questions of philosophical as well as scientific interest, such as whether errors were caused or uncaused, random or predictable, and so on.

Much of this reluctance stemmed from the feeling that there were simply not enough data to warrant speculation. Conferees felt that the knowledge then available did not provide enough sound evidence to support any reasonably comprehensive theory. While such a feeling was understandable, it was (needless to say) unsatisfactory.

Over the next few years there was an upsurge of scientific and public interest in human errors, their causes and their cures. Partly as a consequence of the Columbia Falls conference, the number of scientists in the field increased. The expensive and (to many persons) frightening events at Three Mile Island had made the public more aware that industrial technology had created situations in which relatively few people held enormous power—power to destroy factories, neighborhoods, cities and, in the military context, the whole of mankind. Errors that in the past might have led to the injury or death of one or a few persons could now lead to public catastrophe.

At the time, the study of human reliability was still largely empirical. The efforts were generally directed at improving people's reliability, and not at understanding why they were unreliable in the first place. The amount of basic research, as we have seen, was very small. However, it was clear that there were many more scientists and engineers working in the field—certainly enough to justify a second conference.

A CRUCIAL STUDY: THE BELLAGIO CONFERENCE

We believed then, and believe now, that the topic of human error deserves much more attention than it has received in the scientific literature. It was this belief that led us to suggest to NATO that a specialized workshop be organized at which an investigation of the nature of error could be launched. The NATO Science Committee has as its mandate the encouragement of scientific cooperation between member nations that will improve the social and political well-being of all the NATO countries. The Rockefeller Foundation's Bellagio Conference Center was offered to us as a venue; the Foundation uses it especially for small conferences that further the study of topics central to the social and economic welfare of nations. Twenty-two persons from NATO countries participated in the conference. A list of the attendees is given in Appendix 2.

The Queries

Everybody was asked before the conference to submit a paper responding to a number of topics and queries.
They were asked to *define some terms*:

"error,"
"mistake,"
"fault,"
"slip,"
"accident,"
"cause,"
"reason,"
"origin" and
"responsibility."

There was a series of queries about *theory of error*:

Can there be a general theory of human error or is each error unique?
What is *your* approach to a theory of error?

Are errors ever caused or are they always caused?

Are errors random? What do you mean by your reply?

Is there a common mechanism underlying all errors? A small number of mechanisms? Is there a separate mechanism for each error?

Does it make sense to distinguish between errors arising from internal sources (a lapse of memory, for example) and those arising from external sources (such as poor design of information displays)?

In what sense is a faulty design of a man-machine system a cause of a subsequent error of the actor?

What is an "error theory" a theory of?

Then they were asked to comment on the following questions regarding *prediction of error*:

Can the timing of errors be predicted?

Can the form of errors be predicted?

Does knowledge of the nature of the task, and of the correct action, play a role in predicting errors?

Do you know of any good data on the distribution of the intervals between errors?

Do you know of any good data on error rates?

Do rates vary consistently?

Are there error-prone people?

Then with regard to *"therapy" for error*:

Can error rates be reduced?

Can systems be designed to absorb errors safely?

Can good system design ensure that errors will always be caught?

Is error-free performance possible?

If error-free performance is possible, is it desirable? Would the elimination of error turn people into mere robots?

Can an actor be trained to make fewer errors?

Can an actor try not to make errors? If so, what is the actor doing?

Then some *speculations about error:*

Is there a virtue in error?

Is error an evolutionary survival mechanism?

Could learning occur without error?

Is error needed for adaptation?

Is error related to creativity?

Should people be blamed for their errors?

If so, who: designers, trainers or operators?

Under what circumstances?

The Responses

The responses to these questions were distributed to all the participants before their arrival at the conference. During the conference there were some papers given on specialized topics, and small workgroups were formed to discuss the major topics. Everything was tape-recorded. In addition, each group nominated one of its members to write up a daily summary of the discussions. This book is a synthesis of all the ideas that came out of the conference, including the pre-conference papers, the specialized presentations, the discussions, and the summaries of the discussion groups. All of these were analysed, the elements identified, and all the parts synthesized into a coherent whole.

Some of the questions were remarkably difficult to answer. The definition of error was far more of a challenge than we had expected. To some people "error" was virtually the same as "accident." Others had psychological definitions involving notions of "intention." Some people did not even accept the word "error," as they already developed a theory or model of the phenomenon, and wished to use terms such as "slip" or "mistake" instead. In fact, no general agreement was reached; the participants used words as each saw fit. The problem arises in part from the fact that most of the words we use to discuss research on error already have a meaning in everyday language. Although everybody at the conference generally understood everyone else, there was considerable difference of opinion regarding terms and definitions.

It is important that we, the researchers in error, end this disagreement. Unless we can agree on the use of terms describing research, it will lead to confusion in communicating scientific results. Failure to provide precise terminology could also have a considerable impact on legal arguments about blame and responsibility. One way of dealing with this is to agree on a "technical" meaning or definition of a common term which researchers will use in a more restricted, specialized sense.

The relationship of error to creativity was something many participants preferred to ignore. It had been raised because many people have speculated that there might be a relationship between a tendency to make errors and a tendency to be creative.

In one form or another all the participants responded to these challenges. The following eight chapters present the essence of these responses. Some of the responses were formal and dealt with each of the queries, even those which were repetitive, in turn. Others were essays that touched on some, leaving others unanswered. Everyone responded with something.

ABOUT THIS BOOK

This volume is drawn from the proceedings of a scientific conference. The people who took part in the conference all presented written papers; several of them gave formal presentations to the conference. This book, however, is not a collection of scientific papers identified by author—the traditional form. Most such volumes of "conference proceedings" tend to be very dry and of interest only to specialists in the field. We want this book to reach people in many fields, who may or may not know anything at all about human error. Thus we have presented the material in a continuous narrative, which we hope will be more readable, and more accessible to the non-specialist.

In a very real sense all the people who took part in the conference are the authors of this book; the description of the authorship on the title page is entirely accurate. We as analysts and synthesists have simply tried to present the ideas of the various conferees as clearly as possible. Although we are also represented in the text (because we were participants at the conference) we have done our best not to let *our* ideas and beliefs distort our presentation of other people's. We have not passed critical comment on their views (even where we disagree), but have tried to make a fair presentation of what each of the different participants had to say. The final text of the book was submitted to everyone for approval before being sent to the publisher.

When the participants first arrived at the Bellagio Conference Center, they held a great many views on the issues discussed above. These were expressed in "position papers," which all the conferees had prepared before their arrival. These papers form the basic material for the following chapters.

No point of view is omitted. In those cases where two or more shared some kind of consensus on a particular issue, the conferees' words have been combined to arrive at a clear expression of their views. If an opinion was not shared by any other person, it stands alone. Any quotations in the text are taken directly from the position papers.

This is an analysis and a synthesis. It has some of the flavor of the Book of Proverbs: there is much that is true, much that is thought-provoking, much that is contradictory. There are widely divergent statements which exemplify the differing opinions held by people when they arrived. The last three chapters are based on the opinions expressed at the *conclusion* of the conference, at which point there was considerably more agreement among the participants. These chapters form a section unto themselves.

It is axiomatic that the opposite of an important truth can easily be another important truth. Despite the various contradictions, *every* view expressed here has some claim to validity. It will take a massive research effort to decide which is really true and which false. It may well be that *all* the points of view are true, but for differing circumstances—which would certainly make the research task more difficult, but no less important. It is not enough to say "We need to do research so we can understand error better"; we must also recognize that a more complete understanding of error is not of merely academic interest. It will have profound consequences for human safety and productivity, the attribution of legal responsibility, and the calculation of risk.

As an end to this introduction, let us again make clear what follows. There was a wide range of opinions about most of the topics, whether those of definition, interpretation, research, or practice. In the following sections we have tried to represent all the opinions. To find an opinion expressed does not mean it is correct, or even that the weight of evidence is predominantly in its favor. It means that someone with professional concern for understanding the implication of human error believes it to be, at the very least, a useful or important idea.

Some Queries and Some Definitions

2

DOES ERROR EXIST?

We start by considering an extreme view: There is really *no such thing as error*, and the organizers of the conference were biased in assuming that there exists something that can be called "human error," as if it were an observable object.

> . . . we may observe the performance of an operator, classify it as being incorrect, and determine the cause to be a "Human Error". . . . It is, however, obvious that "human error" does not refer to something observable, in the same sense as decision making does. If we believe that so-called human error is a cause of behavior, not a consequence, it is still true that error must be inferred from observations rather than being observed directly. Error is but one way of describing a human performance and is the term used when no other explanation can be found for a system failure. Behavior still consists of perception, attention, memory, action, etc., all functioning as they usually do. It is only how we classify the *result* that defines the error. Therefore to study error is to study ordinary psychological processes.

This extreme attitude has implications for what could be called the "philosophy of error." Although it is certainly thought provoking, it did not find much support among conferees.

The definition of an error, however, certainly depends on the point of view of the person who judges that an error has occurred. The actor who commits an error recognizes it only after the fact, with the perspective provided by hindsight. Either an actor or an external judge needs a model of task performance to be able to decide whether an action has been correctly executed. We could, therefore, define an error as a human action that fails to meet an implicit or explicit standard. An error occurs when a planned series of actions fails to achieve its desired outcome, and when this failure cannot be attributed to the intervention of some chance occurrence.

Human Control

For an event to be classed as an error, there must have been some possibility of human control. The chess expert sees things that the novice cannot see. The novice therefore does not do things that an expert would do. Since, however, the novice cannot even *conceive* of the expert move, is it entirely correct to say that he or she makes an error? It seems more reasonable to say that where there is no possibility of correct performance there can be no error, even though performance may be imperfect.

One can distinguish between "errors" and "human errors." An "error" is any significant deviation from expectation, depending on statistical criteria or experience of normal performance standards. "Human error" is a deviation from expected human performance, which refers back to the point made earlier that someone judging whether an error has occurred must have a criterion. This particular distinction concerns whether the actor's behavior alone is examined, or the performance of the human-machine system as a whole.

Discrete and Continuous Situations

Typically, in everyday tasks in industry, the operator's job is keeping a variable (temperature or speed, for example) at a particular value or in an acceptable range between values that are too low or too high. In such cases error is generally considered to be present when some value of a variable under control diverges from the desired criterion level.

One possible approach is to think only in terms of the results of behavior. An *error* would be a single malfunction leading to a complete failure of the task with no recovery possible. A *mistake* would be a single malfunction that if not recovered would lead to a complete failure of the task. A *fault* would be a continuous malfunction, with no recovery possible, that causes the task to be performed outside the preselected limits. A *slip* would be a continuous malfunction that if not recovered would result in the task being performed outside the preselected limits. The important distinctions here are whether it is possible to recover from the malfunction and whether the task is discrete or continuous.

To some extent the tendency to consider *all* errors unacceptable may arise from thinking only about certain kinds of tasks, in particular those which are "discrete." Such tasks require responses from the actors which are either right or wrong, allowing no intermediate actions. *This* switch must be moved, not *that* one; *this* meter must be read, *that* decision made. *Any* other action is incorrect. But there are other, "continuous" tasks (steering a vehicle, for instance) where small errors are acceptable, but large ones are not, and there is a range of possibilities from the entirely correct to the totally incorrect. It doesn't matter whether the direction of the car veers slightly from a straight line— as long as it remains in the correct side of the road. Particularly in these classes of behavior, errors have a positive role to play by allowing a range of behavior, all acceptable, during the learning of a skill. Indeed, even when the skill has been fully acquired, small errors act as signals to the driver that an appropriate action is required. In any task where performance is controlled by feedback, errors function as a positive signal for control. Errors can thus be seen as a necessary part of skill development: one must get lost in order to learn the route; one has to make errors to learn skills. Errors as such are always part of control operations that include feedback, and in such situations are neither good or bad. But when errors become too large or badly timed or of the wrong character, they become "grievous errors." A grievous error occurs when the actor exceeds safe operating tolerances, and is what we usually mean in everyday language by an "error."

This view is related to one that regards errors as instances of human-machine or human-task misfit.

When such a misfit exists the results are called "system failures" if they are the result of system variability, or "human errors" if the result of human variability. Errors are unsuccessful experiments in an unkind environment; but only those errors, (in the sense that others would use the term) which lead to an unacceptable consequence need to be studied.

Human or Environment?

In relation to this, there is an important difference between those who attribute an error to the actor and those who attribute it to the system or the work environment. An example of the former: *error* is defined as a divergence between the action actually performed and the action that should have been performed. An example of the latter: *error* is defined as an action or event whose effect is outside specific tolerances required by a particular system. The latter definition does not directly refer to actions of the actor, but only to the consequences of the actor's actions. Even if the actor had intended to do A and instead did B, there would have been no error unless the system in which the actor was working had gone beyond acceptable limits.

An error may imply a deficiency in the actor; a deficiency in the plant should be called a "failure," or "equipment failure." It would be preferable to use "failure" instead of "error" on the grounds that "failure" would imply a context within which the error occurs, and "error" would be reserved as a description only of human action.

Intention

The question of intention is central to the meaning of "error." It seems reasonable to regard "error" as a superordinate category that subsumes all the other words applied to differences between what was intended and what was done; and between what should have been intended and what was intended.

The notion of intention suggests that an actor may have in mind a range of possible strategies with which to approach a task. Some strategies may be less than optimal but sufficiently good in practice to suffice; that is, the effort to try an alternative strategy is not worth it. Thus an ideal performance from a strategic or "economic" point of view may not necessarily be error free. An individual may be prepared to accept some error rate if the workload required to reduce it or eliminate it is felt to be too great.

Intention also means that it is important to separate observable behavior from mental states. "Accident" and "action" refer to observables; "intention" is a mental state and can only be inferred from observation. Words like "error," "mistake," and "slip" require comparison of the internal state and the observable behavior. An error is an unintentional discrepancy from truth or accuracy, a deviation from the expected or prescribed.

Summary: A Generally Accepted Definition of Error

The nearest we can come to a unified summary of the views that were held by the participants prior to the

meeting is to say that for most people error means that
something has been done which was:

not intended by the actor;

not desired by a set of rules or an external observer;
   or

*that led the task or system outside its acceptable
limits.*

However, there are important philosophical issues
involved which must not be overlooked. To at least
one person errors do not exist; for another they are
effects not causes; for yet another only certain kinds of
human acts can ever be in error, since for many acts a
good reason can be given even if to an observer the
action seems wrong. All errors imply a deviation from
intention, expectation, or desirability. A mere failure
is not an error if there had been no plan to accomplish
something in particular; nor is it an error if there was a
correctly conceived plan that simply failed on the
occasion in question.

Finally, errors can be seen in terms of neural events
in the brain, or in terms of psychological mechanisms.
They can be described as sensory or perceptual events,
cognitive events, motor events (that is muscle actions),
actions in a well-defined system control task, or unacceptable consequences in the output of the controlled system. It may be best to call the latter two the
expression and the *consequence* of an error, reserving
the term *error* for an event within the actor.

WHAT IS A FAULT?

By and large, the term "fault" is quite vaguely defined. While it can mean "error," it often carries a pejorative connotation of blame or responsibility beyond that implied by the previous term. It often seems to be used almost as a synonym for "accident"—the product of an error rather than the error itself. For some people "fault" and "mistake" are synonyms; we can include "slip" as well. It is barely distinguishable from "error." It is perhaps better not to use it at all, or to employ it only when referring to problems not in the actor, but in the system with which the actor interacts, as when we say " a fault developed in the heater," or "there was a fault in the computer hardware." Such a use would at least keep a clear distinction between events originating in the human operator and those external to him or her.

WHAT IS A MISTAKE?

There are many definitions of "mistake." A mistake is something the actor intended and which someone else did not intend; a mistake is an incorrect decision or choice or its immediate result, or an error in deciding what is intended. A mistake is an error whose *result* was unintended, as opposed to an *action* that was intended. (For example, if a person makes a mistake in choosing a goal, the action chosen to reach that goal may be correct, but since the goal is incorrect the outcome will not be as desired.)

Overall, a mistake is perhaps most usefully defined as an incorrect intention, an incorrect choice of criterion, or an incorrect value judgment. Viewed thus a mistake can be either "genuine" or "careless." It follows that one may *be* mistaken without *making* a mistake. For example, a sailor may think his compass deviation is 2 degrees east when it is really 2 degrees west, and choose a course which is correct for the former. This would contrast with another conferee's definition of "mistake" as an "incorrect selection of goal."

Although "mistake" has a loose meaning in common language, it has come to have a rather precise technical meaning among those who study error, due to the way it has been used in the work of Reason (1990) and Norman (1988), who distinguish between "slips" and "mistakes." Mistakes are planning failures: errors of judgment, inference or the like, when actions go as planned—but the plan is bad. This use seems now to have become standard terminology.

WHAT IS A SLIP?

A "slip" is defined as an action not in accord with the actor's intention, the result of a good plan but a poor execution. This is in accord with the suggestions of Reason and Norman.

An alternative suggestion is to replace the word "slip" with the term "unintentional error." But this suggestion did not receive much support, since it raises the question of whether any deliberate action, an *intentional* action, should rightly be called an error at all. A possible extension of meaning for "slip" would be to refer to lapses of memory, in line with normal English usage in phrases such as "it slipped my memory." Alternatively, as mentioned earlier, one might tie the word to the question of whether recovery from a malfunction is possible. But right now the Reason-Norman definition is standard. "Slip" refers to an unintended error of execution of a correctly intended action. Some synonyms may be desirable and useful, such as equating "slip" and "lapse" especially in special cases such as referring to errors of memory.

WHAT IS AN ACCIDENT?

An accident involves an unwanted and unwonted exchange of energy. That is, the most important feature is the physical damage involved. Some accidents are the consequence of error; some are not. Fortunately not all errors lead to accidents.

For nearly everyone at the error conference "accident" involved injury to person or property, or carried the connotation of misfortune. "An accident is an unintended event with sad consequences." An accident has unfortunate consequences due to error, mistake, or fault; an accident creates external disturbances interfering with performance, or causes the system to perform abnormally.

Accidents were seen as unpredictable and random. Thus, *accidents happen to us*; they are at least part chance. In general *accidents are not something that people do.*

WHERE DO ERRORS COME FROM?
THE CAUSES, REASONS, AND
ORIGINS OF ERROR

Cause, reason, or origin are all terms that have theoretical implications; they refer to underlying "tendencies for people to make errors." *Cause* suggests the necessary antecedents of an event whereas *origin* suggests the presumed starting point in a causal chain of events.

Do Errors Have Causes?

Errors have "causes" and mistakes have "reasons." There is always some cause for an error although it may not always be illuminating to find it if it was an accident. The term "accident" is used here in the specific medical sense and refers to abnormalities in the brain or peripheral nervous system (such as a stroke or an epileptic attack), as well as gross accidents involving the entire person. However, identifying the cause of system failure with either mechanical or human failure depends on whether one stops at the first or proximal cause or whether one goes back through a thorough analysis that will almost always end up identifying a human error, probably during design or manufacture.

To talk of the cause of errors is to talk about the sense in which they can be explained. Errors and mistakes are not all random: perhaps none are. Errors may be the result of bad design. Errors may be caused by too much or too little automation. So-called "performance shaping factors," which is to say: "anything in the environment that affects human performance," may be causes of error. In general, errors are due to combinations of causes, rarely to a single cause, and this contributes to the difficulty of identifying and preventing them. There are internal and external causes of error.

For example, an accident could happen because of one person's error and another's mistake. Consider a driving collision in which one person thinks the other will stay on the proper side of the road (a mistake) and the other errs, crossing the median. Both error and mistake have causes which can be identified, but it will not make sense to ask for the "cause" of their simultaneity, which is the actual cause of the accident. This situation would hold even for a determinist who would insist on there being some identifiable cause.

There may be a small number of error mechanisms that affect any one individual; there may be an almost infinite number when many individuals are considered. It seems that a search for a cause will always lead back to the "Ultimate Cause" unless we stop at a *practical cause,* which is defined as that point or those points in the causal network about which something can be done.

A slightly different opinion holds that causes of error are rarely discernible even by the actor. It would be foolhardy to attempt to determine *the* reason or cause of an incident. Errors have multiple causes. It is however informative to attempt to determine the major causal factors, because it is these about which one may hope to do something.

One should distinguish between the *cause* and *agent* of error. An actor can sometimes be the agent but not the cause of an error. An example is when an actor is forced to make a decision on the basis of inadequate information. Indeed in such a case one might say that no error occurred. If a decision must be made, the most anyone can do is make it in the light of available evidence. If there was not enough evidence to make the best possible decision, that decision is not necessarily erroneous, even if the outcome is extremely undesirable.

"Cause" is sometimes equated with "reason" in everyday language, and "origin" with "responsibility." On the other hand, "cause" is also equated with "origin," and for that matter may be thought to include "reason." All these terms are somewhat ambiguous and have no widely accepted definition or connotation. It would perhaps be preferable to select a few of them and define them precisely as a preliminary step towards a consistent treatment of the nature of error. There is an extensive literature in philosophy on the distinction between "cause" and "reason," and this is reflected in legal usage as well.

Do Errors Have Origins?

The "origin" of an error might be the same as its "cause," but one could argue that "origin" refers to the *location* of a mechanism, not the mechanism itself. Thus an error might originate in the frontal lobes of the brain, from some unknown and possibly unknowable cause. Or one might say that a pilot failed to notice an overheated engine because his attention was elsewhere. The origin of the error was in the pilot's attention mechanism. That mechanism, however, functioned normally—it was simply misdirected.

The origins of an error are often built-in deficiencies in the design of a human-machine system. These can be the result of incorrect assumptions or omissions on the part of the designer, badly designed user-machine interface, poor training, etc. Thus to ascribe the origin of an accident to "human error" can be seriously misleading if it implies "operator error" when the true origin is poor design.

Is There a Reason for an Error?

In every situation there are a number of factors specific to that situation alone. These are the underlying factors that lead to the error's being more than some residual or trivial probability, and can be spoken of as the "reasons" for an error.

One view was that the reason for an error is the single cause of a unique error. On the other hand one can argue that reasons are grounds for a change of intention; since errors are unintentional, there cannot be any reasons for errors. Mistakes have a more complicated relation to reasons because they have a larger cognitive component. Usually there is not a good reason for a mistake, only rationalizations after the fact. But if a person gives an account of why he or she chose a particular course of action, and the account is sensible in the light of the evidence available to him or her, then the resulting choice of behavior is not an error, even if the results are undesirable. To speak of "reasons" is to speak about choice. Errors are not choices.

The most common definition was that a "reason" is a statement about why something happened; a subjective explanation of goal state or intention. Generally there was agreement that a reason is the psychological explanation of the cause of an event, and that it may or may not be the correct explanation. A reason is a justification that is given for or thought to apply to an action. In some languages there are special words to distinguish between internal and external "causes," which relate closely to this distinction.

SHOULD ACTORS BE HELD RESPONSIBLE FOR THEIR ERRORS?

"To err is human, to forgive divine." In today's world the words of Alexander Pope often go unheeded. We live in societies that are litigious, religious, and unforgiving, and where there is often a tendency to demand that *someone* be blamed for any undesired event, whether accidental or a result of malevolence.

When people at the conference were asked whether actors should be blamed or assigned responsibility for their errors, responses were mixed. There was some agreement that while it is unhelpful to assign blame, people do so anyway. Blame does not generally reduce errors. There was some agreement that blame for error might be useful for certain purposes such as to satisfy the public need for revenge or, as Voltaire put it, "pour encourager les autres." The notion of responsibility should not be abandoned because it can induce guilt in those who have committed errors, and "this constitutes an excellent means of their learning their duty."

The more pragmatic felt that all errors should be reported without expectation of blame, and that only the failure to report should be punished. Although reporting an error may not directly change the probability that the actor will commit another error, "going public" about errors can send strong signals to others, who may then take steps to avoid that particular action. The example commonly cited was the Federal Aviation Administration's scheme in which pilots can report their own errors without punishment. It is widely believed that this has had an important effect on the history of civil aviation safety.

Actors should then be blamed for their errors only if the blaming can lead to improvements in performance, either by increasing the chances of errors being corrected or decreasing the number of errors. We need to discover how effective blame can be in leading to a reduction of error rates. Blaming can be useful if it is part of feedback control of error. Finally, some participants felt that blame is a punishment and should not be expected to be effective in reducing error. That is not its role in relation to error; the difficulties involved in shaping behavior by punishment are well known.

ABOUT SKILL, LEARNING, AND ERROR

Much earlier we mentioned a kind of behavior in which error plays a special role. Error is not necessarily "wrong" behavior as opposed to "right" behavior, but plays a different and constructive role as a control signal. This is the domain of the manual control of continuous dynamic systems such as vehicles. One has to accept errors to develop skill. In trial and error learning wrong routes must be taken to discover right routes. Some errors may be "good errors" that lead to correct solutions.

Obviously errors occur during learning. In fact, our usual criterion for deciding that learning has taken place is whether or not the error rate has been measurably reduced. Thus error appears to be a mechanism integral to learning, one necessary for developing skills in complex tasks. However, it is not the case that the more errors are made during learning, the fewer will be made at a later time.

It may be that learning can occur without error if the learner has not learned anything new but has merely restructured existing knowledge. The principles underlying the Skinnerian "teaching machine" are that error is not only unnecessary but actually an interference in the learning process, and that behavior can be progressively shaped with few or no errors until perfect, fully mature skill is developed. Such a belief clearly contradicts the view expressed in the preceding paragraphs.

Related to the issue of learning and skill is the notion of testing human capabilities to their limits. People take risks and sometimes these translate into accidents. Are these the result of errors or merely of unfortunate concatenation of low probability events and circumstances? In many sports, risk is part of the pleasure, and this is true of life in general for many people. Risk leads almost by definition to error at some time. But this kind of error is, in a sense and in the mind of the participant, not exactly *wrong*, but part of the rules of the game. Perhaps what should worry us is the extent to which people regard crucial tasks as games, and hence do not take error in these tasks seriously (for example, the people who do not take driving seriously, and so feel free to drink alchohol before driving).

ABOUT EVOLUTION

Errors may be an evolutionary survival mechanism at the species level, providing some individuals survive the direct consequences of the errors to benefit from the indirect. This is because an "error" may allow an organism, including a human organism, to explore new ways of behaving which have not in the past been tried, or have been tried and have not paid off. It seems foolish to repeat an unsuccessful piece of behavior. If the demands of the environment are changing, however, then even that which is "known" to be an "error" may still be worth trying again from time to time. Obviously if the negative payoff is *too* great (nuclear warfare is an example that springs unbidden to mind) then there is no virtue in allowing the behavior to occur.

Thus, there may be some optimum error rate. The low rate we see (or think we see) may be the best compromise evolution could make between rigidity and chaos. Consider spiders. Since they rarely make errors, a spider can hardly discover a better way of making a web in its own lifetime. Error is a necessary part of adaptation. It is both a cause and an effect of adaptation.

The evolution of living forms is based on the notion of requisite variety and margin of error. Error is needed for learning, adaptation, creativity, and survival. Slips may usually be destructive, but there may be a virtue in mistakes if they result in learning and their negative consequences can be minimized. Mistakes are the inevitable result of actors creatively experimenting with their environment. Mistakes are necessary to evolution.

ABOUT CREATIVITY

A significant question is whether error is a prerequisite for the existence of creativity. In other words, are creativity and error opposite sides of the same coin?

Trial and error learning is essential to creativity in all fields. Error may function as an adaptive mechanism, as a positive reinforcement in itself. For example, philosophy has been stimulated by the special class of errors called paradoxes, the study of which has led to significant advances in epistemology and logic.

Errors are unexpected behavior. Sometimes such behavior is thought of as being creative, insightful, and indeed valuable. People's behavior often degrades gracefully; the brain mechanisms that produce mistakes of judgment and action as well as slips of performance also allow us to balance the competing demands of life and produce creative behavior. Attention, for example, tends to be diverted easily. This is good because it often leads to novel and creative ideas, albeit at the price of easy distraction and loss of attention to the main task.

SPECULATIONS

One side asks whether there is anything good to say about errors. Are there any moral benefits? It answers: "Probably not. What we should try to do is get rid of them as much and as far as possible." The other side believes that there are many virtues in error, even if most of them are serendipitous. (A commonly cited case is the field of art and aesthetics.) These are certainly desirable, enhancing as they do the richness of our lives. Since it is equally certain that the hazard to which the world is subjected by human error is so great, our major concern must be to understand, predict, and prevent it to the greatest extent possible.

CAN ERRORS BE CONTROLLED?

Is a person who errs responsible for the error? A possible answer is that if there is no evident cause or if it cannot be found, then the error was a mental "act of God" and not something for which the actor was responsible. More generally "I am not responsible for my errors but you are for yours"—the main reason to discuss responsibility is in order to ascribe blame, often in a legal context. On the other hand novices are expected to make errors as are the tired and the overworked (but these latter are often not forgiven their errors). Often the punishment for an error may not be obviously proportionate to the gravity of the error. Consider for example in the way society appears to regard medical errors. If a patient is subjected to an incorrect procedure, the surgeon may be punished more severely than if the patient dies on the operating table.

In another frame of reference, responsibility is seen as a matter for jurisprudence and should be left to the courts to decide.   It falls into ethical and judicial frameworks, and is not the province of science.

Responsibility is also seen by many as being a way of pointing to a kind of cause. Thus assigning responsibility is a way of indicating which aspect of a poorly defined user-machine system caused an accident to happen.   Responsibility is the quality of being the prime causal agent for an event: that is, something that forms the conditions giving rise to behavior not in accord with the goals of the system. If I tell you to carry out a task in a particular way, and this results in an unacceptable performance or even an accident, I am *responsible* for the outcome, whether or not you made an error. If you *do* make an error, I am *responsible* for it even though I did not *cause* it.

"Cause," "reason," "origin," and "responsibility" are steps in a backward process of analysis after an accident.   These terms are used in confusing ways. "Cause" is closest to the event, and usually refers to mechanism(s) that are necessary precursors to some malfunction or other event. "Origin" is a description or claim of how internal mental functions failed (human malfunction, discrimination, input information and so on). "Reason" would be found in an analysis of factors affecting human performance, such as subjective goals and intentions. "Responsibility" must generally be limited to a use in a legal frame of reference, where there is a clear need for such a term.

GENERAL COMMENTS

There was more agreement among the participants than at the first conference about the definitions or connotations of each of the words about which they were queried. Complete agreement about any of the topics was never achieved. Even the word "error" itself is still somewhat ambiguous although there was more agreement on it than on anything else. If our understanding of error is to progress, there needs very much to be an effort to define some terms carefully so as to avoid common usage and cover everything which needs to be named. The nearest thing to an example of what is required at present is the agreement on the distinction between "mistakes" and "slips" proposed by Reason and Norman.[1]

It seems that one can make out a case that *error* should be the superordinate term and that particular kinds of error should be named by adjectives appended to the word "error." Investigators in different laboratories and different countries have very different ideas about what to call things. The confusion that can result is more important than a mere difficulty in scientific communication. It is particularly important if work on human error is to be clearly communicated to the public and to professions such as engineering, industry, the law, and government.

---

[1] In the work of Reason and Norman published since this conference a mistake is defined as an incorrect intention, and a slip as an incorrect and inadvertent action. These usages have become standard.

On Taxonomic Issues 3

ON THE NATURE OF TAXONOMIES

In any area of scientific study, it is necessary to develop a clear-cut system of classification. Unless everyone can agree on what is being studied or discussed on a particular occasion, there is no point to study or discussion at all. Such a system of classification is called a *taxonomy*. The conference participants were asked for their opinions as to the most fruitful and appropriate taxonomy for human error. It turned out that there was no agreement about a single taxonomy which would serve for all purposes of error research.

What Criteria Must We Use In Classifying Errors?

In general, a successful taxonomy must be related to both the theoretical and practical purposes of the investigator. In the study of error, there are a number of major categories of research, each with its own appropriate taxonomy. In the case of sensory or motor errors, for example, one could in principle have a

41

neurophysiological theory, together with a taxonomy
in terms of neural events. On the other hand, if
mistakes (defined above as "errors of intention") were
the category of interest, then we would need a cognitive theory. Since we do not presently understand the
neural substrate of intention or thought, forming a
taxonomy on that basis would lead to insurmountable
problems.

When we examine the varied phenomena of error,
and the many ways in which error has an impact on
human life, it is clear that there are many directions
that research in the field could take. Each area of
research may need its own taxonomy. In fact, one of
the most striking features of the conference was the
large number of taxonomies proposed to classify what
is a rather small amount of research. There is an
important relation between how one does research
(what measurements are taken, what research designs
are used, etc.) and the taxonomy chosen. We must be
clear at what level we are operating. Some data permit
only the most superficial level of classification. In
general, categories at one level do not neatly map onto
other levels.

Most of the taxonomies suggested at the conference
involve several levels of analysis (*what* happened, *how*
did it happen, and *why* did it happen?). The taxonomies themselves may thus be classified according to the
level at which they approach the problem of classifying
errors. So what follows is a taxonomy of taxonomies;
a listing of the ways in which we can organize the data
available to us.

There Is More Than One Way to Classify an Error.

a. First, there are *phenomenological taxonomies.* These describe errors superficially, with terms that refer almost directly to the events as they were observed. Typical categories in such taxonomies include "omissions," "substitutions," "unnecessary repetitions," etc. In applied areas, where the emphasis is on interaction with machines, classes such as *recoverability, the attribution of error either to human or machine,* and the *nature of the consequences of the error* are common.

b. At the next level come taxonomies of the *cognitive mechanisms* involved. Errors are classified according to the stages of human information processing at which they occur. Thus they are divided into errors of *perception*, errors of *memory*, errors of *attention*, etc. This is the level of taxonomy that perhaps is most strongly supported by the great mass of research literature from modern experimental psychology.

c. The next level includes those taxonomies which assign errors to classes based on the *biases or deep-rooted tendencies* they are thought to reveal. For example, one might classify an error as due to the tendency of humans to form a hypothesis about what is happening, and then seek only evidence to support that hypothesis, rather

than testing it—the "confirmation bias." An
early version was Bacon's "Idols of the tribe."[1]

d. Almost the only point of agreement at the conference was that it is fruitless to look to classifications based on *neurological events*.

The need for a variety of taxonomies seems to be felt
because of the different purposes that the participants
had in mind, or the different tasks with which they were
concerned. For example, there has been a major
attempt to develop systematic classifications of errors
for the purpose of estimating human reliability in the
nuclear industry (Swain and Guttmann, 1983). This
has given rise to a taxonomy based on the task itself:
what is the probability of misreading one of three
gauges, of pressing one button when it is intended to
press the neighboring one, etc. A taxonomy aimed at
nuclear plants includes a specification of the task, the
workplace, the crew, the work environment and so on.
It is a way of classifying what we would call the *ex-
pressions of error* rather than the errors themselves. It
is essentially a phenomenological approach.

One approach used factor analysis to try to discover
the minimum number of categories needed to classify

---

[1]More recent examples are the "representativeness, availability,
and anchoring" heuristics of Kahneman, Slovic, and Tversky (1982), the
hindsight and overconfidence biases identified by Fischhoff (1975) and
his colleagues, the "illusory correlations" of the Chapmans (1967), the
"confirmation bias" that crops up in many guises but especially in the
work on reasoning by Wason and Johnson-Laird (1972), and schematic
distortions in memory.   These latter were first identified by Bartlett
(1943).

all the errors that were observed in his research. It was found that the factors needed to account for errors were different from those describing correct performance. There are errors in working with *numerical* material, with *figural* material, and a third factor representing *"hasty and inadequate responses."* This represents an attempt to let the psychology of error, as it were, speak for itself in defining the way in which to classify errors. This level, however, is probably too coarse for many purposes, making it necessary to look at subordinate classes of error, such as those described in other taxonomies.

One relatively extreme view was that there is no need for a taxonomy of error as such. Behavior which is in error is first and foremost *behavior*: hence what is required is simply the same taxonomy of behavior that is sufficient for describing normal (error free) behavior, plus a taxonomy of tasks and hardware to put the behavior in context. This view follows from one that feels no need to define error in the first place.

Another respondent did not think there can or should be a theory of human error. From that perspective, therefore, a taxonomy should be concerned with the situations where "mismatches" can be observed rather than with the inferred "human errors." There are many possible taxonomies: in one sense the "mismatch" can be attributed to the machine, to the man, or to the interaction between the two; it could be attributed to an external agent (e.g., a supervisor in a nuclear power plant). There are others. The choice of a taxonomy depends on the purpose of classification. However, error is not necessarily just variation of performance. In hindsight, less successful variants of performance are called errors. A sufficiently good model of the world, one that contained enough detail to show what features of the environment caused the observed behavior, would render the concept of error unnecessary. Most participants, however, saw error taxonomies as useful and indeed necessary, however simple they might be.

TWO PROPOSED TAXONOMIES

At least two of the taxonomies proposed were in fact taxonomies of behavior. The first of these is Rasmussen's taxonomy. The second is Moray's modification of Altman's taxonomy (Altman, 1966). Their proponents saw them as a way of classifying errors both in terms of the psychological mechanisms involved in their production and of the stages of information processing underlying the errors, not as a way of getting rid of the concept of error.

Rasmussen's system includes 6 general categories and 31 specific categories, which at first seems very elaborate. On the other hand, the taxonomy was developed to analyze data from very complex industrial systems. It is a mistake to oversimplify for the sake of elegance or convenience when the phenomena suggest that complexity is indeed required. The main categories of Rasmussen's taxonomy are:

1. observation of system state;
2. choice of hypothesis;
3. testing of hypothesis;

4. choice of goal;
5. choice of procedure;
6. execution of procedure.

The taxonomy, despite (or perhaps because of) its detail, is not really appropriate for many everyday tasks, such as, for example, rote domestic work. Rasmussen, however, states that the use of a coarse or a fine-grained classification system depends on whether you are working on theory or design. As already stated, this taxonomy was developed in a particular industrial context of very great complexity.

Moray suggested a modified form of a taxonomy originally proposed by Altman. In this, class of error is related to psychological mechanisms. It classifies errors by their level of behavioral complexity, their mode, the type of learning involved, and the relevant psychological data from other research to understand origins of error. The levels of behavioral complexity are:

1. Sensing, detecting, identifying, classifying;
2. Rote sequencing;
3. Estimation with discrete or continuous responding;
4. Logical manipulation, rule using, decision-making, problem solving.

For each of these levels there are modes of error and types of learning and training involved, as well as relevant psychological theory, data, and models.

A TAXONOMY OF
PSYCHOLOGICAL MECHANISMS

The other major scheme that has been extensively used for research in recent years is that of Reason (1990). His taxonomy is at the intermediate level, level b (see p. 43), which uses *psychological mechanisms*. These include mechanisms such as *strong associate substitutions* (in which actions strongly related to the same memory may be substituted for one another); *place-losing* (in which steps of a sequence of action may be omitted or in the wrong order); and *interference*. The first cause absent-minded slips of action; the second omissions, reversals or repetitions; and the third, blends or spoonerisms (a sort of "cross talk" between simultaneous information processing activities).

Reason attributes some of these *level-b* phenomena to *level-c* mechanisms. He discusses constancy mechanisms, spatio-temporal partitioning biases, ecological constraints, built-in limits in processing, schematic biases, heuristics, and the fact that the brain seems to be designed to detect change rather than steady states of stimulation.

Like Rasmussen's and Moray's taxonomies of behavior, Reason's taxonomy seems on the fringe of being able to predict the occurrence of errors. Those task characteristics that are liable to cause "strong associations" or other such phenomena, would be likely points for an error to occur. While this does not unequivocally guarantee prediction, it is at least a step in the right direction. Psychological theory should be able to identify such features of tasks.

SOME OTHER BASIC
CLASSIFICATIONS

The number of categories in proposed taxonomies varied enormously. At one end of the continuum are simple binary taxonomies. For example, Senders pointed out that errors may arise within the actor or within the environment, and that therefore a simple basic classification is in terms of *endogenous* and *exogenous* errors respectively. Even such a simple scheme has useful corollaries. For example, if an error is endogenous, that would suggest training or motivation to be an appropriate way to reduce it, while an exogenous source of error suggests system redesign.

A particularly popular binary taxonomy distinguishes between *slips* and *mistakes* as forms of error. These, unlike the binary class mentioned above, are related to the actor and not the system with which the actor is interacting. Slips are errors of execution; mistakes errors in the formation of intention; and, slips are low level errors, a by-product of skill; mistakes are high level errors. Skill brings efficiencies to work but only over a limited range of situations. It is also possible to devise taxonomies within the two classes of the slips-mistakes binary classification, based on the presumed causes of the slips and/or mistakes.

Several other relatively simple taxonomies were proposed. Thus a distinction was proposed between *formal human error* as a transgression of rules, regulations, and algorithms, or an out of sequence performance; *incoherent human error* as a nonrequired performance; that is, some output not stimulated by a system relevant input; and *substantive human error* as an unintended performance, for example, because the procedure was inadequately defined. A four-fold system was suggested based on the distinction between *correct* and *incorrect possibilities* in the performance of a defined task and *approved* and *disapproved ways* of performing voluntary actions.

There were also several references made to earlier attempts to classify errors, for example that by Kollarits (1937), using substitutions, omissions, repetitions, and insertions, and that of Bartlett (1943). The latter specifically discussed cognitive errors thus: "(They) are often: a. errors of *omission*; b. errors of *timing*; c. errors of the '*third kind*,' solving the wrong problem; and d. *attention failures*, tunnel vision, disintegration of visual field." (Bartlett, 1943).

Underlying a taxonomy of *slips* and *mistakes* is the following hypothetical mechanism: when people intend to do something, they will think about their intended actions. In this process, an organized unit of knowledge, called a *schema*, is activated. Each schema is triggered by certain conditions which can therefore be called *trigger conditions*. When the trigger conditions for a certain schema occur, that unit of knowledge will influence the actor's performance. An error can occur when one of several events occurs. The wrong schema may be activated; there may be an error in activation; or there may be an error in triggering. These form the 3 categories of slip: *capture*, *mode,* and *description* errors.

There was some feeling that the simple binary classification into "mistakes" and "slips," though convenient, was too coarse to be of much practical use. But that criticism is met by the more elaborate version outlined here and described in more detail in the appendix. A more problematic issue is the extent to which the taxonomy is useful only for analysis of errors after the event; while it is valuable for research and theory building, it cannot be used to predict the imminent occurrence of errors.

SUMMARY

Many taxonomies may exist. Depending on the task to be performed and the nature of the research, a particular set of classifications will be preferred. Starting from a given set of data, we might describe the very same phenomena, the very same errors, in several different ways. We can have errors of *perception*, *intention,* and *execution.* When the emphasis is on communicating to non-specialists, we can classify errors of *omission*, *substitution*, *insertion*, and *repetition*, and we can use the classes *exogenous* and *endogenous* so others will know whether we lay the errors on design or on human misfunctioning.

As already mentioned, the need for refinements that depend on the purpose of research or discussion was emphasized by a comment that Norman's classes of slips and mistakes, mentioned in the previous chapter, are "too coarse" a taxonomy "to be operationally useful in guiding the design of systems and training programs." Probably the most general conclusion from the presentations prepared prior to the conference is that there is a need for an eclectic variety of taxonomies which will depend upon the particular task being analyzed. These may range from global classes such as errors of theory, strategy, tactics, and response, down to the most refined and detailed analysis of small "atoms" of behavior. At present no single picture is dominant.

On Theories of Error 4

IS THERE A THEORY OF ERROR?

A theory of error is certainly possible. Most of us would consider it necessary. If we have a theory of error and the consequent understanding of its mechanisms, we should be able to minimize the occurrence of errors and mitigate their consequences. It is therefore rather strange that the participants in the conference proposed very little systematic theory in presenting their ideas about error.

Opinion was divided on the extent to which errors are unique, and require unique explanations. There would probably be agreement that most errors have some characteristics and mechanisms in common but that each has some unique characteristics. Since the definition of an error depends on the point of view of the observer, the latter will affect the theoretical approach seen as relevant to the explanation of a particular error.

IS AN "ERROR THEORY"
DIFFERENT FROM A THEORY OF
BEHAVIOR?

As we mentioned previously, it was suggested that the very idea of "errors" as objects of distinct study was incorrect, and that a more accurate perspective is that errors are simply behavior that is undesirable in a particular context. This view implies that a study of error is therefore nothing more than the study of behavior and its contexts. Although that viewpoint was not widely accepted, it may be that it was implicit in people's minds when they approached the idea of a theory of error.

If so, a theory of error should be included in a theory of behavior. Error is treated as an integral component of normal behavior, and if we do not have a good theory of behavior, the related theory of error can not be any good either. It is clear that a particular behavior is correct in a given context , and incorrect in another. With a good theory of behavior, we can form an understanding of "error" by discussing the situation in which it takes place. The "theory of error" would then be a theory of why a particular piece of behavior occurred in a particular context at a particular time. The study of error would be identical to the study of psychology.

A "SPECIAL THEORY" OF ERROR?

If on the other hand we believe that a special theory of error is desirable, what properties should it have? A theory of errors should describe, predict, or account for behavior in relation to what the ideal behavior in that situation would have been. In other words, it should be *normative*. That is, it should include a theory of physical performance plus a theory of value for relevant events. As we have noted, theories interact strongly with taxonomies, and the theory we work with will depend on (and use terms drawn from) the appropriate taxonomy.

SOME DIFFERENT APPROACHES
TO A THEORY OF ERROR

One could have a theory of causes and a theory of reasons. The *causal theory* would link chains of contingent events and the *reason theory* would deal with the justification of actions and the assignment of responsibility and blame. One may need to distinguish between endogenous and exogenous errors, between those caused by events within the person and those caused (despite the actor's best efforts) by the environment. An important question is whether endogenous errors are caused by different mechanisms from exogenous errors and therefore require a different theory. Although there was not complete agreement on the point, the distinction between these two types of error appeared to be a natural one. Indeed, it was claimed by some that "from practical experience . . . most errors are exogenous in the sense that the external . . . factors are usually the most important in generating errors." Situations where this is not true are usually the result of poor quality of training and work practice (as, for example, badly designed shift-work schedules).

Should errors be defined and explained in terms of *outcomes*, or in terms of *processes*? To adopt the former position would lead to theories about overt behavior and observable events in the performance of the human-machine system. (This is commonly the approach taken by reliability engineers in probabilistic risk assessment.) To adopt the latter would mean that we need a theory couched in terms of psychological mechanisms and their properties. Even within a theory of processes it may be necessary to distinguish levels of processes or levels of theory. Lower level processes bring efficiencies to behavior but only over a limited range of situations, and often by making use of unconscious, habitual actions. Slips are low level by-products of skill; mistakes are high level, the results of information processing at relatively "high" levels.

If one thinks of skilled behavior, for example, there are two levels at which one might talk about its origins. A high level explanation would be in terms of what the actor decided to do after thinking about the situation. Errors at that level would require a theory of mistakes. On the other hand, a "lower" level of explanation might be appropriate for slips: the actor did not intend the action at all, but some property of the neuromuscular system (which controls movements at a level below consciousness) produced the action. The theory might even be couched at this level in terms of the biomechanics of movement rather than psychology.

In some cases a higher level of theory may be required, one at which personal or social values are the determining factors in causing errors, or at least describing events as errors. It may be, for example, that each person has a preferred level of risk at which he or she likes to operate. Rather than trying to reduce the risk of error to zero, each actor tries to behave so as to keep behavior in as efficient a state as possible with a constant rate of error. One might call this "risk homeostasis," by analogy with the self-regulating homeostatic mechanisms of the body which act to keep body temperature, chemistry, etc., at a constant preferred value. According to the theory of "risk homeostasis," one cannot reduce error rates or guarantee to design error-free systems of operators and machines. Operators will increase the risks they will take as the overall system becomes less error-prone. The only solution is to make the *social* consequences more rather than less serious. In other words, make errors into "sins!"

Another point of view is that a world without sin and a world without error are equally "silly" and cannot exist. It is necessary to consider both behavioristic and humanistic theories of error to bring about a better world. Clearly at this level of discussion we are approaching problems of the relation of blame and responsibility to error, and are invoking psychological constructs such as motives, social meaning, etc. It is very unlikely that a coherent theory of error could be developed at present at this level, especially if the aim is to predict the occurrence of error and design steps to reduce it. Somewhere on this scale from low to high levels of theory lies the theory of Freud as described in "The Psychopathology of Everyday Life." Adherence to Freud's view was conspicuously absent from the conference.

SUMMARY

In summary, at the outset of the conference there was very little in the way of formal theory of error proposed by any of the participants, although theories of particular classes of errors (such as the control of skilled movements or sensory judgements) were suggested. The latter were, for the most part, standard psychological theories or models of perception, attention, movement, etc., and were not specifically theories of error as such.

On the Prediction of Error 5

At the heart of our interest in error is the question of prediction. It is certainly a matter of intellectual interest to speculate, after the fact, as to why a particular accident occurred, how human error contributed to it, and what were the psychological and other mechanisms that caused it. If, however, we are condemned by our ignorance merely to sit weeping amid the ruins, the study of error is simply self-indulgence. On the other hand, if we can determine when and where an error will occur, and who will commit it, then there is at least the possibility of preventing it, or of responding more rapidly and effectively to it when it occurs. This, from a practical point of view, is *the* question about error: can it be predicted? If exact prediction is not possible, can we predict the probabilities of error? This would still be of great value.

HOW CAN WE PREDICT ERRORS?

Errors can be defined only in relation to correct and desired behavior. It follows that for any situation, a knowledge of the correct action is essential for both prediction and prevention. Also crucial are the details of the task configuration, or (at a more general level) a complete logical analysis of the system design. To understand and predict errors, we must understand comprehensively the properties, purpose and operation of the human-machine system. This usually requires a detailed task analysis.

The choice of a taxonomy can influence the ease or possibility of prediction. A very specific taxonomy implies a high level of detail. While this is useful for purposes of definition, it actually interferes with our ability to predict error. It is possible to make a very general statement, for example, that errors are more likely between 3 a.m. and 4 a.m. on the night shift than between 3 p.m. and 4 p.m. on the day shift. But it is extremely difficult (if not impossible) to say exactly when a *particular* error, or any error at all, will occur.

It is easy to feel despair at the problem of investigating something as elusive and qualitative as human error. But one should not be too overwhelmed by the presence of uncertainty in one's estimates. Even in physics one gets approximate answers. Thermodynamics and Heisenberg's Uncertainty Principle both argue against strong predictive power at the level of fine detail. At the gross level, however, one can do very well indeed. Error theory is likely always to include statistical uncertainty because, ultimately, error depends on the functioning of the central nervous system. There are too many factors affecting people's behavior to allow for exact, deterministic predictability.

PREDICTING TIME AND FORM

Errors result from the normal operation of the human information-processing system, along with effects arising from the environment, the various pressures and biases influencing the actor, and the latter's mental, emotional, and attentional states (ignoring the possibility of traumatic events that damage the functioning of the nervous system). In principle, if we knew all these factors, we could predict errors precisely. In practice, since we cannot know all the factors, we will always have to resort to statistical prediction.

This is true both of the form and timing of errors. Because there are so many possible causes of error, and the relationship between these is very complex, we cannot predict timing with any degree of precision. In very simple tasks, the timing of errors might be predicted fairly exactly, but in realistic tasks timing cannot be predicted (except in the mythical case when an error *always* occurs with a given interface design and a particular operator and circumstance). Obviously, the timing of errors could be predicted exactly if we had control over *all* the antecedent conditions. Equally obviously, once we are out of the laboratory, we have no such control.

If a system is used for a specific task by a number of different people, a variety of errors will occur, which will probably differ from one person to another. We may be able to predict the timing and form of errors by theoretical or statistical analysis of the appropriate learning curves of a human being. A careful analysis of the errors that people make during the time when they are learning a task should provide us with clues to the errors they are most likely to make when proficient, and under what conditions these errors will arise.

PREDICTING PROBABILITIES

Considerable effort has been expended in the past few years in estimating the probability of human error for a wide range of actions. Many tables and handbooks have recently appeared, but there is considerable skepticism about the accuracy of their estimates. By and large, one can make only quite general statements about possible types of errors, about their frequencies and their comparative likelihoods.

We should remember to distinguish between *error rates* and *point probabilities*. The former are steady state probabilities, the latter the probability that something will happen on a particular occasion. Error rates are frequently expressed in ratios. For example, one may say that the error rate for misreading dials is one misreading, on the average, for every thousand times that the dial is read. Some estimates of error rates have been made and vary from actions that are almost always incorrect to as few as one error per 10,000 opportunities.

Most of the published data on error rates have wide uncertainty bounds placed around them. These are often ignored by users who don't follow statistics, who simply need numbers for decision making. Those who have collected the data note the importance of "performance shaping factors," which can cause wide variations in error rate. (For example, it is well known that the probability of error is very strongly affected by circadian rhythms, which cause a very large increase in error rates in the small hours of the morning.) The few experiments which have been performed suggest that error rates are highly unstable. One study found that the probability of an error varied from .01 to .95 over 250 trials. On the other hand, some measures at a rather global level seem to be fairly orderly. In diagnosing nuclear power plant faults many data show that the time to achieve a correct response is predictable.[1]

The point probability of an error, on the other hand, is likely to be a different number from the steady-state probability. Some of the time it may be an entirely different concept. Are the two concepts and the two values the same? This is likely to be a theoretical question; both one's theory of behavior and one's theory of probability will dictate the answer. Good data on error point probabilities are very difficult to collect, except in very carefully controlled, very well simulated situations and tasks.

We have already seen that there is considerable difficulty in defining error. If we can not do this adequately, making predictions or estimates of the probability that a specific error (or any error) will occur is much more difficult, if not impossible. Our definition of error may depend on the nature of specific

---

[1] It is a "log-normal" curve. The probability that the correct response will have been made by the team of operators is proportional to the logarithm of the time since the fault occurred.

situations, tasks, and machines. If so, our prediction
will be equally restricted. Furthermore (just as we
need a definition of error for our estimation of probabilities) it is also necessary to decide just what constitutes an "opportunity for error," an *occasion on which
the error might have occurred.* Since the "steady-state
probability" of an error is the ratio of the number of
times it *did* occur to the number of times it *might have*
occurred, defining these latter occasions will have a
profound effect on our calculations.

The available estimates of error rates are probably
artificially consistent because they are subjective. They
are often based on the estimates of experts, rarely on
real data collected from real tasks or even simulators.
Probably, when we have good data, there will be still
less consistency. It must also be remembered that since
operators detect many of their own errors and correct
them before they have an observable effect on task or
system performance, the true error rate is always greater
than that which is observed. (Whether or not this is felt
to be important depends on whether one is interested
in practical consequences of error or in the theory of
error.)

Although error rates can be predicted in principle
(and sometimes in practice), nobody is really sure just
how reliable these estimates are. To some extent, one
can also predict changes in error rate. It is important
here to note that very often predictions using terms
such as "greater" or "less" ("ordinal" measurement)
rather than estimates with exact numbers may be
enough. For example, in introducing modifications to
a system with currently acceptable error rates, it is
quite enough to predict that "the errors will not in-

crease," without stating their exact probability. It is reasonable to expect that as more data are collected we will be able to predict better what error rates will be in any situation. For some special cases we can make strong, ordinal predictions already. For example, the frequency of writing the date of the prior year on cheques is very high in January.

PSYCHOLOGICAL BASES OF ERROR

If a proper regimen of training and testing has been implemented, we should be able to predict whether a human operator will *know what to do*. But this does not mean we can predict on a particular occasion whether he or she will *actually do it*. There are many variables in the operator's physical and mental state that make such a prediction impossible. Furthermore, in many cases tasks are performed continuously or repeatedly over a long period of time, and by teams whose composition varies from shift to shift or from day to day. Since we do not always know who will be doing a task, we cannot be sure how much knowledge about the system will be available on any occasion. Even when we have a good idea of the psychological processes involved in the task, we are still faced with this additional uncertainty. Therefore we can only turn to predictive models based on probability, such as Markov Process models, which predict error rates in tasks where people, such as maintenance workers, work at their own pace.

It can also be argued that error rate data for human beings do not adequately describe behavior in the case of some errors. Such data as we have mentioned earlier are based on observable events, such as slips of action. But in many cases internal, mental errors, such as mistakes (to use Norman's term) are the major problem. Slips can be predicted only on a general statistical level, that is, they are *probabilistic*. Some mistakes, on the other hand, are deliberate cognitive decisions by the actor; the resulting incorrect actions stem from these decisions. In a sense, these actions are deterministic with respect to the *time* they occur, but (paradoxically) this does not help us to predict them. Both slips and mistakes are probabilistic if we try to anticipate the *form* they will take: *what* error will occur, not *when* it will occur. (Contrary views in respect of the prediction of form are presented later.)

The prediction of mistakes is different from the prediction of slips. If we can determine the information-processing requirements of a particular task, *and* assess the completeness, accuracy, and availability of this information (either from a specific human's knowledge or from the system displays) then we can predict with certainty whether or not mistakes will occur. Basically, if the information available to the operator is defective, there will be a mistake. An erroneous action that follows from a mistaken decision is neither a slip nor an error. Given an initial mistake, the action itself is correct if it follows logically from the decision that was made—"It seemed a good idea at the time." We need to know the probability of the mistaken decision even more than the probability of the "incorrect" actions stemming from it.

Another position is that we can predict both the *forms* and the *relative rates* of error, if we have a clear understanding of the details of the task and the circumstances in which it is to be performed. This claim stems from the degree to which the various subtasks have common elements. The forms of errors can be predicted with situation models and adequate identification and quantification of failure modes. To accomplish this, both task analysis and a finely detailed examination of the design of the workplace are necessary.

When we examine data from simulations and field experience, we can conclude that after an error occurs, *other* problems will develop and persist. Often they are not corrected by the actor who is making the error; in some cases an incorrect assessment of the situation leads to the directing of attention to the wrong displays, and so on. Having more time does not always mean a better solution to the difficulty; there is often no speed-accuracy trade-off. The persistence of errors in such situations arises because low-level action sequences are performed without conscious thought. To rectify the situation, the actor must return to higher levels of information processing where the evidence can be reassessed. Another problem (and a particularly important one) is "confirmation bias," the tendency for people to be biased in favor of their original hypotheses and diagnoses.

Error Proneness

The concept of "error proneness" has had a rather checkered history. Error proneness may simply be a statistical anomaly. If we observe someone when their error rate is high (due entirely to random variation) and not at some subsequent time when their error rate is low (again due to random variation) we might incorrectly describe the person as "error prone."

There do appear to be trait-like variations in different actors' error rates. Broadbent, Baddeley, and Reason have each used questionnaires to find out about minor cognitive failures, mostly in memory and attention. The data suggest that people do differ in enduring ways. There is a small but consistent relation between error rates on the one hand and accident and near-miss rates on the other. The evidence further suggests that people who make one kind of cognitive error, for example, errors of attention, also tend to make other kinds of cognitive errors, such as errors of memory. This may imply a common underlying cause for many different kinds of error. Broadbent attributes this to the way in which people deploy a universal but limited mental capacity, probably attention. A high rate of minor errors reflects a chronic problem in the deployment of attention. He also finds that individuals with a high error-rate are more vulnerable to stress.

On the other hand it is obvious to anyone who has investigated the reliability of the human operator that the frequency and occurrence of "human errors" depend less on any stable inherent characteristic of the operator than on his or her interaction with the environment. It is not enough to know even the error rates characterizing a particular person—we need detailed descriptions of the situations where design mismatches occur.

Research on inter-error intervals suggests that many errors are random in time. We know that error rates follow a circadian rhythm; many people believe that they are more likely to make errors at some times than at others. Many data exist, however, which suggest that over short periods when the actor is in a particular state, errors occur at random intervals. For example, while more errors occur in the early hours of the morning than at other times, this model suggests that the moments when errors occur *within a particular one-hour period* would be randomly distributed. It appears also to be the case that some people make more errors than other people. If this is our definition of error-proneness, then it exists. There is much less evidence to support the claim that there exists a group of individuals whose error rates are so high that they must be regarded as pathologically different from the general population.

The way the intervals between errors are distributed is also relevant to the question of error-proneness. If an error is made that destabilizes a system, so that it becomes more difficult to control, then there will be a greater chance of another error following immediately. Later, as control is regained, there may be a reduction in error. Surprisingly, while such phenomena are seen sometimes in laboratory experiments, there are no good data for "real" tasks, only anecdotes.

Paradoxically, even if error-proneness exists, it may not be a serious problem in many complex real industrial tasks where prolonged training is required. Chronically error-prone persons would not stay in real systems; they would get fired. There may be brief increases in error rate resulting from fatigue, stress, or other factors, but it would be wrong to consider this evidence for error-proneness.

Personality Correlates

Many suggestions have been made about the relation of personality to error. For example, it has been speculated that low self-esteem might increase the probability of accidental self-damage, and that anger might lead to a higher probability of aggressive behavior and consequently harm-causing actions. It has even been suggested that there may be an unconscious use of errors to gain tactical advantage, a view similar to Freud's proposal that errors are the result of unconscious motives. There seems little to be gained from such approaches, and in our present state of knowledge, behavioral scientists might just as well ignore them. However, these ideas may be of value in describing systems of voluntary actions (like driving a car) where other, purely behavioral and mechanistic theories, have been notably unsuccessful. The predictive capability of personality tests is too small to have any major impact on the prediction of errors at the moment-to-moment level.

On the Reduction or Elimination of Error 6

HOW CAN WE REDUCE ERROR?

The study of error is basically an applied study. We would like to develop ways of reducing or eliminating error, of reducing or eliminating the impact of error-producing situations on people and the environment.

Our knowledge of the causal mechanisms of error is limited. As we analyze the factors that lead to the choice of a specific action, we will be able to specify the correct decision process for a given situation, system, and state. We can reduce error rates by altering the human-machine system in a way that depends on this analysis. We need to understand better the nature of cognitive tasks, and to construct models of human-machine interfaces based on our knowledge of cognitive processes. This is clearly an important area for research, and one where there has so far been relatively little progress.

Since we are concerned with human-machine systems, we can approach the reduction of error either through engineering or through psychology. But we must not forget that the ultimate solution is a *systems* solution, in which the requirements of engineering and of psychology will both be satisfied. (By "engineering," we mean both the traditional disciplines of mechanical, electrical, and civil engineering, as well as the newer disciplines of cybernetics, computer science, etc.)

REDUCING ERROR THROUGH ENGINEERING

Let us assume that we are dealing with a system that is prone to malfunctions, including those caused by "human error." After we do an analysis of the system, we may approach the reduction of error through engineering. We can redesign the human-machine interface, or redesign the system so that its behavior is changed in the appropriate ways. If it is impossible to eliminate the error-producing behavior completely, we can redesign other parts of the system so that it becomes "forgiving," absorbing errors without letting them lead to catastrophic results. In some cases we may be able to alter the role of the operator, or design hardware that supports his or her actions and compensates for his or her failures. Finally, there may be a case for designing the human beings out of the system altogether, by increasing the role of automation, and perhaps including artificial intelligence.

Traditionally trained engineers will frequently pursue the last alternative, choosing to remove the human operator. This can be dangerous in ways that are not immediately evident. The operators of complex systems are usually highly trained and highly educated, especially relative to the qualifications of the maintenance and support personnel. When we increase the degree of automation, and design the operator out of the system, the maintenance crew becomes more and more the main source of errors in everyday operation. Furthermore, if there is a design flaw concealed within the system, removing the operators robs us of a highly flexible and responsive line of defense against hazard.

Operators are not the only ones who make errors. Many may be made at a higher level—supervisory or even management. Such errors are very difficult to catch. They remain latent until particular conditions of the human-machine system occur.

If we use the *human factors* (psychological) approach, we are concerned with improving human performance. We can select new operators who are less likely to make errors than the current personnel, improve training so that everybody's level of performance will be improved, motivate the operators to perform better, or change the social consequences of making an error.

Design For Error Reduction

Even if we choose an engineering solution, it is essential that we take human factors into account. A properly designed system must take into account the properties of the people who use it. There should be no "mismatches" between the cognitive, psychological, social, and physiological characteristics of the human operators and the design and response characteristics of the system.

Systems can be designed to absorb errors. It is obvious, though, that we can not design a system to respond to errors that we cannot anticipate! It would be best to predict forms or modes of error, so as to design in advance for their absorption. Since in many cases errors are "close to" the desired or intended behavior, either functionally or "geographically," (i.e., two adjacent switches) it will often be relatively easy to predict the errors that can be expected. In other cases we may be able to invoke psychological data, such as what are called "population stereotypes." For example, in many countries, a switch that is in the UP position is OFF. In North America it is ON. We can predict that if we mix foreign and North American equipment, there will be errors that stem from the users' nationalities. By extension, one should be able to design systems so that the errors that will occur are likely to have insignificant consequences. From time to time bizarre and incredible errors will be made, however, and the *real* challenge is to design for these.

Redundancy in System Design

A common approach to improving the reliability of multi-component mechanisms is to introduce redundancy. If a component's probability of failure is very high, we can use two or more, in parallel or in series. The probability of their all failing simultaneously is very much less than the probability that any one of them will fail. This principle has also been applied to the improvement of human reliability, with the use of two or more actors working in parallel or series as appropriate. The improvements in reliability , however, are much smaller than is the case for machines. Common-cause human failures—failures which affect both actors in the same way at the same time—will probably vitiate the effects of redundancy despite the best efforts of the designer. Alternatively one could have operators check one another, and have computers check the operators, as well. But, wherever we have several people involved, there are complex social dynamics due to role and status which make it extremely difficult to tell what will happen. If one person checks the work of another, the result should be an increase in reliability—unless each becomes careless, relying on the other to catch errors.

We can also use another kind of redundancy, saying things to the operators in several different ways. This could be built into the format of displays, but might be more effective if used to monitor the execution of operator commands. When the operator receives or reads a command to execute an action, it would not be carried out immediately; confirmation would be requested. It might be desirable (despite the necessary delays) to display to the operator a prediction of a particular command's result, and to request approval. Only if the operator approved the request in the light of the prediction would it be carried out. It may be easier to monitor errors than to train human operators to operate optimally. A computer could, as it were, "whisper suggestions in an operator's ear" and suggest that an error has been made.

PSYCHOLOGY AND TRAINING IN
ERROR REDUCTION

If we rely on training and psychology to reduce error, we must start by adopting a different attitude towards errors. One can regard error only as a damaging phenomenon, or as a clue to the processes in the central nervous system that are responsible for the behavior. As far as the origin of the behavior is concerned, there is no psychological difference between a trivial error and one with disastrous consequences. The difference lies in the circumstances in which the behavior occurs, and in how those circumstances affect its impact on the environment.

All the usual methods of industrial psychology will work. Personnel selection, training, warning, exposure control, and so on can all reduce error rates. Actors should be provided with "knowledge of results" (KOR) at all levels, including design, construction, installation, operation, maintenance, and management. KOR must be immediate, to the point, gentle, and supportive. No one should be "put down," intimidated, frightened, or confused by this feedback. Its purpose is to give the actor a signal that can be used to modify behavior. Motivation can be created and maintained by good system design, which means not only the human-machine interface, but the working policies as well. People need to be challenged and motivated by the use of simulator training and other methods that go beyond the usual textbook procedures. These could include classroom debriefing and discussion, quality circles, improved job design, etc.

As far as possible, people who are less likely to make

errors should be selected for hazardous or economi-
cally important tasks. Personnel selection, however, is
not a very powerful method of improving systems.
Another way to get error-free human machine systems
is to integrate design and training. If operators show
signs of becoming fatigued during operation, the machines should be redesigned. If the machines show
signs of producing bad parts, the operators should stop
stop using them. Machines should be "error loud"
(give alarms) and operators "error proud" (intolerant
of errors).

In some tasks the level of training is already extremely
high—landing on an aircraft carrier deck, for example.
The improvement to be gained from more training is
slight. For most tasks the rate of learning has slowed,
after two or three hundred hours practice, to the point
where further training does not result in any substantial improvement. It must be emphasized that people
can be trained only to perform better, not necessarily
to make fewer errors. An emphasis on the elimination
of one error may lead to an increase in other errors as
the pattern of behavior shifts, and this shift may be
unpredictable. Learning is a dual process: acquisition
of the required responses and inhibition of competing
ones. Error is a necessary part of learning.

Proficiency and Error

It is important to distinguish between the errors of the
beginner and those of the skilled operator. The errors
of the novice arise from incompetence and have a large
random component; those of the expert stem from
misplaced competence. Since they tend to take the

form of well-established, well-practiced cognitive routines, they are often predictable. They are what the human operator has *learned to do*. The greater the proficiency, the less the probability of mistakes, but the greater the chance of making such a "skilled slip," because the skilled operator pays less deliberate and conscious attention to what he is doing. It is all the more important to understand the mechanisms of error production. One needs a good theory of human action to know how to design systems in which errors will be prevented, corrected, or absorbed.

IS ERROR REDUCTION POSSIBLE?

It may be that despite our efforts we can not do anything about errors. Actors trying to make fewer errors may succeed in doing so in one area at the expense of increased errors in another. This may come about because they change their strategies and procedures (which could lead to new types of error) or because their evaluation of the desirability of various outcomes (the "payoff matrix") has changed, making primary concerns of secondary ones and vice versa.

Humans are generally very proficient in detecting their own errors of action. Good design can help the actor catch those errors which will inevitably be made. Perhaps as long as we can detect and correct errors, there is no need to do anything about them. There is some evidence, however, that suggests that while people are good at catching their own errors of action, they are much less good at catching their own errors of thinking, decision making, and perception.

If error-free behavior is non-creative, do creative people make more errors? Our casual observation indicates this is true. There is some evidence that the personality characteristics of error makers and of creative persons are similar. Errors are probably a natural development of evolution, a means by which creatures can explore alternatives. At the very least, errors may be the price they pay for that exploration. But should we therefore eliminate creative people from the operation of human-machine systems?

Complete elimination of human error is as impossible as complete elimination of machine failure; there will always be an irreducible residual. Performance that is nearly error-free, on the other hand, is certainly possible. Everyone makes an error from time to time; one who did not would be considered a machine, not a person. "To be error free is to be an automaton." On the other hand "absence of errors would turn people into Gods, not robots." "To be error free is to be sin free" (and highly unlikely!).

On the Proper Classification of Error 7

A TEMPORARY DEFINITION OF "ERROR"

Even simple questions about the nature of error can be very difficult. What is the best way to classify errors? (is there a *"best"* way?) For that matter, what *is* an error? At the conference, one of the discussion groups suggested the following definition, which we shall use for the purposes of discussion: *If there is general agreement that an actor, Z, should have done other than what Z did, Z has committed an error.* Note that this may include internal acts that are only indirectly observable, (such as errors of perception, errors of mental arithmetic, or errors of understanding.)

WHY IS A TAXONOMY NECESSARY?

As we have noted throughout this book, the description and classification of errors is important for several reasons. There is an intimate relation between the way errors are classified, the way their occurrence is explained, and what can be done to reduce their frequency or their consequences; we shall go more deeply into this relationship in the following pages. Another more basic reason is that a taxonomy is a fundamental requirement for the foundation of an empirical science. If we want a deep understanding of the nature, origins, and causes of human error, it is necessary to have an unambiguous classification scheme for describing the phenomena we are studying.

There is no such generally accepted taxonomy, and furthermore, it seems unlikely that there will be one. There may be almost as many taxonomic schemes as there are people interested in the study of error! That is not to say that we can not agree about the nature of errors when they occur, but rather that there are several taxonomies, each of which is applicable to a particular occurrence of error.

Why should this be so? Let us consider some of the purposes that may motivate the study of error. A lawyer will require a taxonomy that can be used to ascribe responsibility for an error to a person or persons. The defense or prosecution of someone accused of negligence may depend strongly on the nature and genesis of a particular error .

Alternatively, the designer of a human-machine system will need to incorporate knowledge of human reliability and probabilistic risk assessment. In this context, the focus is on the probability of error of various types and the mechanisms causing or giving rise to the error. Here, the aim may be to identify aspects of system design, human performance, or workload demands that predispose operators to make errors. It may be necessary to trace the fine detail of those actions in operating a system that can lead to desired or undesired outcomes. If the designer assigns probabilities to each of these elemental actions, it is then possible to calculate the overall reliability of the entire system. Those actions leading to undesirable outcomes can then be used to implement changes in system design or operating procedures that may reduce the probability of significant error.

Academic researchers may wish to categorize errors in order to understand behavior and to develop a model of the mechanism or mechanisms involved. Applied psychologists and engineers may wish to understand error behavior in order to design new human-machine systems that will be less prone to produce error and more resistant to its consequences.

Faced with such a diversity of purposes, it is not surprising to find such a diversity of taxonomies. It turns out, however, that there is a systematic relation between these different approaches. This leads to what may be called a "taxonomy of taxonomies." From this meta-taxonomy users may pick the system of classification most appropriate to their needs, with confidence that it will at least be consistent with those chosen by others, in a logical, causal relationship. It is important to note that we are concerned with *errors*, not with *accidents*. It is probable that were it not for their undesirable outcomes, there would be little interest in errors except as a facet of the scientific study of behavior. Error without undesirable consequences may be considered by most people as merely an irrelevant nuisance.

HOW MUST ERRORS BE
DESCRIBED?

We start with the need to describe errors in terms of *actions* and *interactions*. This level of description, a "phenomenological taxonomy," involves categories such as errors of *commission* or *omission*, and could refer to the design and operation of equipment—controls, displays, etc. This level is centered on the *observable manifestations* of errors—what the actor did or did not do.

A deeper level of description is in terms of the internal processes, the information processing or cognitive mechanisms, real or hypothetical, involved in the error. These processes may be such things as *diagnosis, decision making, hypothesis formation, activation, choice of tactics*, etc. This level acts as a sort of interface between the phenomenological description of "external" errors and the psychological mechanisms underlying them.

These fundamental psychological mechanisms themselves form the next level of description and classification. One can divide errors in terms of the mechanisms controlling the goal directed behaviors. Such mechanisms are *perception, decision, attention, distraction, available response choices, capacity limitations*, and the like.

Finally, we can return to the external environment in which the error occurred, treating it now, however, as cause rather than effect. Errors may be related to general features of the environment that were relevant in a particular case or situation—*glare, noise, distracting telephone calls, social pressure, stress,* etc., or to specific features such as *design of equipment, adequacy of available information, ease of identification and use of controls, illumination*, and so on.

This sequence of taxonomies provides descriptions of links in a causal chain running from the environment, through the person, and back to the manifestation of the error in the environment.

Which one should be used? It depends on the situation and the problem. When *questions of reliability* are at issue, a taxonomy at the first level is adequate to compile data. To *predict or understand errors of decision* a taxonomy at the level of information processing is more suitable. If *redesign of a human-machine system* is the issue, then a taxonomy dealing with environmental and other external demands on the operator is appropriate.

It is interesting to note that one can start from several different points. One traditional means of developing taxonomies has been the application of factor analysis to data. In theory the "natural" taxonomy will emerge from the data; the whole concept of taxonomy is based on the idea that there is a natural and discoverable orderliness in the environment. But the results of a factor analysis themselves need interpretation. The factors are statistical clusterings that point to common sources of variance. Identifying these clusterings and matching them with phenomena and mechanisms calls for judgement by the statistical analyst, and thus to a great extent depends on the concepts and models the analyst already possesses. In short, the analyst's expectations of possible phenomena and mechanisms greatly influence the development of a taxonomy.

SOME TAXONOMIES
ALREADY IN USE

In the last twenty years, almost a dozen proposed taxonomies of error have been published. Several appear earlier in this book. In some way most follow the general outline we have presented. For example, it is common to distinguish between errors caused by events in the environment from those caused by events in the nervous system or the mind. This distinction is sometimes labeled, by Senders for example, as the distinction between *exogenous* and *endogenous* errors—those which have their origin, respectively, *outside* or *inside* the person.

This distinction is useful and widely applied; it is important to keep in mind, however, that no commitment to a *particular* theory of the body-mind dichotomy is intended here. Rather, we sometimes find it convenient to talk in terms of psychological constructs, at other times in terms of neurophysiology. Thus in the chapter on causality we discuss an error caused by the blindness that occurs in a migraine attack. We can use neurophysiological terms, and describe the *neural events* that cause the scotoma in the visual field—(events in the body) or keep to a description in terms of a *failure of perception* (a mental account).

Most schemes include a list of types of error at the first level mentioned above. For example, it is common to find lists such as errors of omission, commission, intrusion, repetition, substitution, sequencing, etc. Among those using such distinctions are Kollarits, Norman, Reason, Altman, and Moray. These are all categories of error distinguished by differences between actual and desired behavior. Norman and Reason both give clear accounts of the various levels at which tactics and mechanisms of information processing are involved—mistakes, slips, schemata, activation, habits, and so on.

The use of the category *"mistakes"* to include incorrect plans, intentions, or goals appears in a number of different ways. Sheridan, for example, considers them part of an evaluation function. He considers "error" a sub-set of all the things that can possibly happen when intention becomes action, action becomes observable behavior, and behavior influences the state of the system. Each step in this sequence is subject to evaluation by "judges," which are hypothetical internal entities. His formulation particularly emphasizes the "closed loop" nature of errors: the effects of actions include subsequent effects on perception, decision making, and future evaluation. By contrast, Norman and Reason concentrate more on "open loop" action generation, and tend to ignore input, or information acquisition errors. McRuer points out that the importance of the consequences of error may be an important distinction. In certain activities, particularly continuous closed-loop control situations such as vehicle guidance or laboratory tracking tasks, there is almost always some error that is used as a control signal, indicating to the controller (whether human or automatic) what control action to take. In general, he asserts, such "error" is not the sort of error for which any theory is required because it is an inherent characteristic of closed-loop behavior. What we need to explain are unusual or severe errors, which are behaviors qualitatively different from normal control errors in such tasks.

In some cases it is not clear whether it is correct to

speak of an "error" at all. If someone misreads a dial because of glare, is that really an error? The operator looked at the right instrument, attempted to read it, processed the available information as well as possible, but did not perceive the actual value displayed on the dial. If the observer extracts all possible information present, then even if the result does not accord with the objective value that is displayed, why should the result be regarded as human error, even if it is wrong? Statistical decision theory indicates that for a given ratio of signal to noise there is a limit to the accuracy with which perceptual judgements can be made. With a low enough ratio there will be substantial numbers of missed signals, mis-identified signals, and false positives, (that is, reports of signals that did not occur). These are undoubtedly errors, but, given that better performance is not even theoretically possible, there is a sense in which they are not human errors: the actor cannot avoid them. The appropriate taxonomic distinction would seem to be that they are *exogenous human errors*. It is interesting to compare the distinction between exogenous and endogenous errors with that of the thirteenth century philosophers, between "actus hominis," an act that emanates from a human but for which he is not responsible, and "actus humanis" a human act deliberately done by an individual. The parallel is very close.

THE REQUIREMENTS OF A
TAXONOMY OF ERROR

In the English language, we tend to lump all undesired actions by humans under the word "error." In order to distinguish between those errors for which the actor is and is not responsible, further refinements in vocabulary are required. Certain languages presume the distinction made in mediaeval Latin. French, for example, distinguishes between "defaillance humaine" and "erreur humaine." It is important that whatever taxonomy is adopted should be transportable not merely between laboratories or industries but across international and linguistic boundaries.

There is a tendency to concentrate on individual behavior, on why *this* individual at *this* moment makes *this* error, in building models and theories of error. Such an emphasis is reasonable when the primary concern is theory-building. It is not particularly helpful when we are examining error in other contexts such as the legal ascription of responsibility or a systems approach to design. It may even be misleading. The very fact that in user-machine system design there is an increasing emphasis on the systems approach demonstrates the way in which causality and responsibility are distributed among a variety of agents, including designers, trainers, managers, operators, and maintainers. This point will become particularly clear when we examine the relation between error and accident in the chapter on Causality. For now it is clear that a taxonomy must be broad enough, diverse enough, to encompass answers to a great variety of questions, not merely those concerned with the psychological mechanisms that cause errors.

A TAXONOMY OF TAXONOMIES

Applied research into error is almost always in the context of accidents. Investigators wish to know what caused the accident. Was it the result of human behavior? Was it the result of deficient design? Could it have been avoided by better training or by personnel selection? Was the operator an active or passive participant? What role did environmental and social factors play? Was the actor inefficient in the use of the information displayed? Were the displays inefficient? Did social or personality factors affect the actor? Was the actor under time pressure or emotional stress? An examination of these and related queries suggests that they can be grouped into four major kinds of questions, and that the groups can form the basis for a taxonomy of taxonomies. From such a starting point we can proceed to a framework with a very general applicability. This structure will preserve a common language, yet allow investigators the flexibility to use particular taxonomies for particular purposes. Furthermore, it will also allow investigators to continue to use their existing taxonomies, yet permit easier translation between taxonomies. The following three tables present the scheme.

Table 1 proposes a classification of the purposes of a taxonomy.

Table 2 shows the organization of levels of taxonomic

TABLE 1:
The Different Purposes of a Taxonomy

THE PURPOSE:	BY WHOM USED:
The assignment of responsibility	Lawyers, courts
Design and assessment of systems	Engineers, designers
Understanding of human behavior	Behavioral scientists
Understanding of psycho- pharmacological agents	Medical scientists
and so on.	

TABLE 2:
The Different Levels of a Taxonomy:

THE LEVEL:	EXAMPLES:
Phenomenological	Omission, substitutions
Hypothetical internal processes	Capture, overload, decisions
Neuro-psychological mechanisms	Forgetting, stress, attention
External processes	Poor equipment design

description. Four levels are proposed. At each level a variety of particular taxonomies will be permissible. But the concepts, the logical type of explanation, will be constant.

Finally, Table 3 lists four levels of query which might be asked about any error.

Each question suggests a particular kind of explanation and a particular range of phenomena which are relevant to the questions.

THE FOUR LEVELS OF INQUIRY
ABOUT ERROR

The first two levels of questions, *"why did the error occur?"* and *"what error occurred?"* relate strictly to *errors*. It would be possible to ask "Why did the driver of the automobile make a wrong turn and go off the road?", and to receive the answer, "because of the fog," which looks like a level 4 answer. Although the fog *could* be the cause, it *must* be an indirect one. No fog is thick enough to change the direction of travel of an automobile. The causal chain leads through the properties of the visual system to the psychological

TABLE 3:
A Taxonomy of Taxonomies

LEVELS OF QUERY:	USED BY WHOM:	FOR WHAT PURPOSE:
1) Why did error occur?	Psychologists, lawyers, courts	To assign personal responsibility; to advance understanding of behavior; to develop methods of eliminating error.
2) What error occurred?	Applied scientists, reliability analysts	To develop estimates of differential error rate; to design systems resistant to error; to identify sources of system unreliability.
3) Which object was involved?	App. scientists, designers, reliability analysts	To improve design of work environment to assign equipment responsibility; to assess system reliability.
4) To whom and where and when? did it occur?	Lawyers, courts, behavioral scientists, reliability analysts	To assign responsibility; to increase understanding of human behavior; to increase understanding of effects of stress, fatigue, narcotics, work schedules, etc.

mechanisms of perception and decision making and so to the driver's error. We discover that "why" must be restricted to questions about neural or psychological mechanism.

These levels, particularly the first, refer to generic events, such as perception, reading text, attention, pressing buttons, and the like. In order to understand what has happened, the analyst does not need specific details of the environmental setting in which the error occurred. The two lower levels are more task specific: they refer to the particular task and the particular setting in which the error occurred. It should, however, be noted that from another viewpoint the first and the last levels, *why* and *where* (or *when*) are generic (the first for the classification of endogenous causes and the latter for the classification of exogenous causes), while the inner two, *what* and *which*, are specific. The "levels" are not really "deeper" or "shallower" than one another. Rather they are pointing to logical dependencies. If there were no system there would be no task, if no task no action, if no action no nervous system to produce it. In order to clarify a particular error usefully, the important point is to identify the appropriate level of description.

Statements made at any level may provide causal answers about an error: "because of the fog;" "because he was driving;" "because of incorrect steering;" "because of misperceiving the shape of the road." We shall see in the chapter on Causality that none of those is *the* cause: the causal chain stops where it is most useful to stop it. But having a clear picture of the different classes of taxonomy can prevent confusion and disputes about what is claimed in an analysis.

Note also that this scheme does not replace existing schemes, theories, or models. The models of Norman and Reason are appropriate to one level. McRuer's distinction between errors that are part of performance (and for which no explanation is therefore needed) and others that are "grievous errors" falls into a different level because it is behaviorally oriented. It is clear that the Altman-Moray scheme needs to be divided into more than one level, because it classifies errors both behaviorally and according to the mechanisms that generate them.

A further important point is that the scheme explicitly points both to the actor as the source of *endogenous* error, and to the environment in its broadest sense as a source of *exogenous* error. In that way the tendency to blame the actor for being careless or not trying hard enough is offset by the indication of causes as remote as the work of designers and the policies of managers.

We recommend that all investigations of error should explicitly include the use of the four classes of taxonomy as a descriptive system prior to deeper analysis. Effectively the scheme provides a check list that makes clear both to the user and subsequent readers what kind of analysis is being undertaken.

THE NATURAL HISTORY OF ERRORS

The relation of the taxonomy of taxonomies to the natural history of errors can be seen by expanding Tables 1, 2, and 3 and combining them into Figure 1. In order to bring out clearly these relations, we shall analyze a hypothetical situation in which an industrial

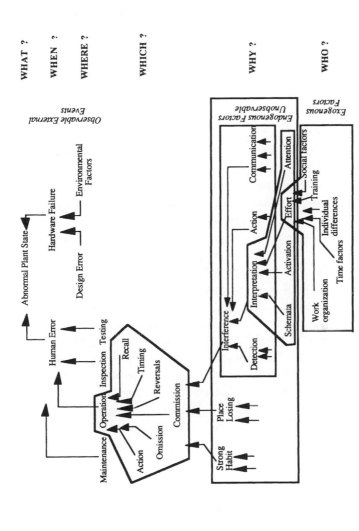

FIG. 1. Causal Network for Error Analysis. Each factor at a higher level has a similar network of causes feeding into it from below. In this example we emphasise the effect of exogenous causes on effort, of effort and other factors on interpretation, of interpretation and other factors as causes of interference, and of interference as a cause of commission errors in operating the plant. Many other causal chains could have been chosen.

plant that is at least partly controlled by a human operator has entered an abnormal state of operation that may result in an accident.

Before giving a general account of Figure 1, note the components in the top center which end in "Hardware Failure." Hardware failures occur, and may lead *directly* to an abnormal plant state (as when a pipe breaks or a boiler bursts), or *indirectly*, as when an instrument or sensor fails, in turn leading to human error because of the loss of information by the operator. Hardware failures themselves may be the result of direct environmental challenge (lightning, earthquake, fire), or be due to proximate human error (improper maintenance), or remote human error (a flaw in design or management policy). In a fully automatic robotic system it would seem that only errors in maintenance, design, or management will have an impact on the system. *Any of those activities, maintenance or design or management, could be an entry in the taxonomy of tasks,* even though they are remote in time and place from the abnormal plant state. Thus the genesis of error is not necessarily an "open-loop" phenomenon in which the task demands call forth erroneous behavior from the operator. In at least some cases a human error remote in time and place causes a local, current human error; *the causal chain of successive human errors propagates through design.*

CLOSED-LOOP ASPECTS OF ERROR GENERATION

At two more points the taxonomy of taxonomies emphasizes that at least certain aspects of error generation are "closed-loop." First, psychosocial factors are part of the *"where* and *when"* data, and appear in addition as answers to the *"who"* question. In all these levels they are accessible to the analyst. Second, the demands generated by the abnormal environmental state become inputs to the mechanisms and factors described in the *"who* and *why"* taxonomies.

Although the diagram is intended as a description of taxonomies, it is clear that in drawing it up we are implicitly describing causal connections. And the fact that there is a closed loop emphasizes that taxonomic causes are not "levels" in the sense that there is some ultimate deepest level of description which gives *the* cause, the *real* cause, of error.

For clarity, many links in Figure 1 have been omitted. In principle, all tasks receive an input from all effects, and effects from all the relevant psychological mechanisms. The latter in turn are all dependent on the basic information-processing mechanisms and underlying neurophysiological structures of the brain (although we seldom need to describe at that level).

When the problem of describing errors and classifying them is viewed in the framework of the taxonomy of taxonomies, many practical implications of research are readily apparent that seem much less obvious in the absence of such a framework.

A PRACTICAL APPLICATION

We shall consider a practical problem for analysis and demonstration. Suppose that a new computer-based control system is installed, and it is found that there are a large number of errors due to the operator's omitting an operation in the start-up sequence. Could we eliminate this error by going to fully automatic start-up? Only if there were complete reliability in hardware components (hardware environmental factors) and maintenance; above all, only through proper design. If there is *any* doubt that the design engineer can design the operator out for that particular error, and keep the plant operational in all other respects, then the operator *must* be left in. The fact that the error occurs frequently suggests that it may be due to a faulty display or procedure. If so, Figure 1 suggests that it should also occur in regular operation, maintenance testing, inspection, etc. If it turns out to occur mainly or only in start-up, this narrows the possible causal tree. If no commission or out of-sequence errors, etc., occur with the same task, the causal tree is further pruned. If the analyst were now to discover that only certain operators made the error it would suggest analysis at the *"who"* level rather than at the *"why"* level.

Other similar chains of reasoning will occur to the reader. As has been stated several times in this book, there is an intimate relation between causality and taxonomic categories.

In concluding this chapter, we should once again emphasize that what is proposed is not a new theory of error. All existing theories of errors from the statistical to the psychoanalytic can be seen to occupy an appropriate place in Figure 1. By analogy with biological taxonomy, Figure 1 is a tree of phyla, families, genera, and species, by which a particular specimen of error can be identified and related to its ecological niche. It differs from a biological taxonomy in its closed-loop causal property. It is a way to integrate and organize any existing and future theories of error.

On the Causes of Error 8

DO ERRORS HAPPEN BY CHANCE, OR CAN WE IDENTIFY THEIR CAUSES?

Earlier in this book we presented a diverse set of explanations for errors, which we derived from the participants' responses to questions posed prior to the meeting. In this chapter we shall examine what the final positions and opinions were after much discussion had taken place. Whether we explain errors in terms of information-processing models such as those of Norman and Reason, or we invoke the theories of depth psychology, which are currently much less popular, all the accounts assume that a cause or a number of causes can be found for any error that a person makes. Let us examine this claim.

It is natural to assume that errors in general have causes, and that each particular error has a particular cause. Humans seem to prefer seeing natural history as chains of causal rather than stochastic events; people prefer to believe in determinism at least pragmatically. We must, however, remember that there are events in the natural history of the world that appear to be uncaused, such as some subatomic (quantum mechanical) phenomena.

PREDICTING ERRORS

If we conclude that particular errors are random events, that their times of occurrence (their "dates," as it were) cannot be foreseen, it still seems clear that the *type* of error is predictable to some extent. The type of error may be determined exogenously or endogenously. To put it crudely, a bicyclist cannot "press the accelerator instead of the clutch," as the environment makes such an error impossible: there is neither accelerator nor clutch on a bicycle. This is a case of exogenous determinism. A person doing mental arithmetic is unlikely to make a mistake in vocal pitch, since the nature of the mental activity is not such as to produce musical sounds; this is a case of endogenous determinism. There are intermediate cases, such as errors in steering a boat or a taxiing aircraft. In both cases the possiblities are confined by the vehicles to errors of directional change in horizontal motion (hence are exogenous), but the steering movements are determined by habits derived from automobile driving, steering with rudder pedals, or steering with a tiller, and these are endogenous causes. Errors are not random in the sense that all possible actions or thoughts are equally probable in any situation.

But if errors are caused, why are they unpredictable? There is general agreement that the best that we can hope for is probabilistic prediction, the kind that is used for human reliability estimates in probabilistic risk assessment. We can at least in principle predict the probable rates of occurrence of errors, using phrases like "errors per 1000 operations," or "errors per hour." We can make at least ordinal judgement as to whether error rates will increase or decrease. We may even be able to say that it is almost certain that *if* an error is made, it will occur in a particular short period of time, or in a particular place, or while carrying out a particular task with a particular piece of equipment. These predictions remain stochastic in nature, however, and no one believes that in the foreseeable future it will be possible to pinpoint the exact moment at which a particular error or any error will occur.

There seems no reason to believe that there is a recognizable state of the nervous system that occurs just before an error is made. Indeed a logical analysis of the various meanings of "error" strongly suggests that if there is such a state it could only be associated with a small subset of errors, such as "slips." In cases involving interaction between people, it will never be possible in principle to predict the error of an individual unless we know precisely what the other person involved in the transaction will do.

Even if we assume a recognizable "error-prone state" in the brain, it could not logically exist in *all* cases where we may wish to speak of error occurring. It is conceivable that just before a "slip," a particular pattern of neural activity indicates that the impending action is the result of random bursts of neurons, and not controlled patterns initiated from the voluntary motor cortex. If we found such activity, we could say the error was purely endogenous and entirely independent of the system in which the person was working. The particular error that occurred, though, would still be highly system-specific. In other words, we might be able to observe the nervous system and say "in the next second an error will occur," and because of the control panel design be fairly sure it will be an error of control action (rather than, say, misreading a display).

In some situations an undesired outcome can not be considered the result of an "error state," even though it results from human action. Consider the case in which an operator uses a shortcut instead of following the standard operating procedures. Since it saves time, the operator does not consider this an error. If the result is a system failure, it will be considered "due to operator error." Since the operator considers the choice of behavior (using the shortcut) correct, it cannot be ascribed to an "error state"—by definition a neurological "error state" cannot be used by its owner to produce "correct" behavior. The fact that currently we are unable even to predict errors probabilistically with any great success is due more to a lack of data than to the underlying nature of error.

PROBLEMS OF CAUSALITY

At first the question of causality seems absurd. We know perfectly well that there are any number of "causes" of errors. Alcohol in the blood "causes" driving errors. Oversleeping "causes" a missed appointment. Lack of practice "causes" a pianist to make mistakes when playing. Similarly, training reduces errors, directions prevent one from getting lost, and arranging to be called prevents oversleeping. Apparently, there are causes for errors, and by identifying them we can reduce the chances of their occurring.

On closer inspection the picture is by no means clear, and difficulties arise in many areas. In the first place the very concept of "cause" is notoriously difficult to understand. For over two thousand years philosophers have tried to define clearly what it means to say that one event causes another. A typist hits an incorrect key. What is the cause of the error? Is it bad positioning at the keyboard? Misreading the text? Fatigue in the muscles? Distraction by the noise of a door slamming? Anger at the boss which brings about a "motivated slip?" Even if it is the anger, in what sense is *that* the cause? Is it that anger changes the way in which the brain sends signals to the arm muscles? Is distraction "really" competition between two parts of the brain for access to limited processing resources? Are not electrochemical events in nerve fibres the "real" cause of arm movements? On the other hand, could we not also say that the error is not caused by the typist at all, but by the boss, whose behavior results in the anger and hence in the movement error?

It is obvious there is a great deal at stake here. If we cannot identify the causes of errors, how can we prevent them? There is, furthermore, another aspect of the problem. As we have seen, although it may often be possible in practice to identify a plausible cause of an error, it is not usually possible to say when it will occur. We can identify situations that tend to increase the probability or frequency of errors, but we cannot predict the exact moment when the error will occur. Indeed, some data were quoted to show that errors occur "at random" as if a purely chance mechanism was at work (Senders & Sellen, 1987) in the same way that radioactive atoms change state at random. If errors occur at random, in what sense can we predict or control them?

ERRORS AND ACCIDENTS: AN IMPORTANT DISTINCTION

The situation is complicated by the tendency often to think of accidents and errors as one and the same. In this book we are concerned with *errors*, not *accidents*. "An accident may be a manifestation of the consequence of an expression of an error." Alternatively, an accident can and frequently does happen when no error has been made, or where the error was so remote that it cannot sensibly be considered the cause. If a person is killed by a falling tree that is struck by lighting when the person is trying to get as far from tall trees as possible, no error is involved. If a rock falls off a cliff, strikes an automobile's wheel rim and bursts the tire, making the car leave the road, no error is involved (it is unreasonable to say the designer was erroneous in not making the tire thicker).

CAUSAL CHAINS

Figures 2, 3, and 4 provide examples of "causal chains" that lead to errors, and in some cases to accidents. They will repay careful examination! In each example the arrows show the direction of causality.

In these examples there are several events that can

Causes and Reasons
a: Typing Error

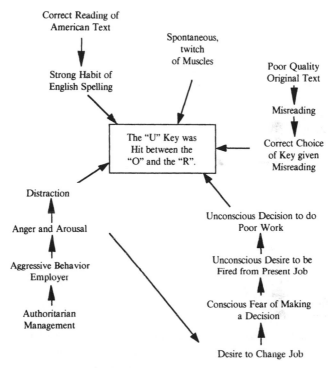

FIG. 2. Why did the typist type "colour" not "color"?

plausibly be considered to "cause" the outcome. Perhaps it is more accurate to say that there are several *causal chains* leading to the outcome, since it is always possible to work backwards in what appears to be an

Causes and Reasons
b: Vehicle Collision

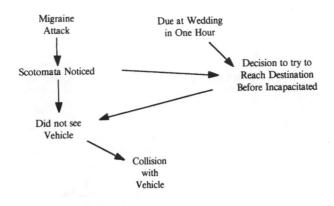

What is the Error?
What is its Cause?
What is its Reason?

FIG. 3. Why did the collision occur?

infinite regression from the outcome to more and more
distant causes. It seems best, at least from a pragmatic
point of view, to say that what is deemed to be *the* cause
of an accident or error depends on the purpose of the
inquiry. *There is no absolute cause.*

If one were interested in assessing skill level, the
cause of the typist's error might be considered an
incorrect finger movement. If one were interested in
minimizing keying errors by designing a better key-

Causes and Reasons
c: DC-10 Crash in Chicago

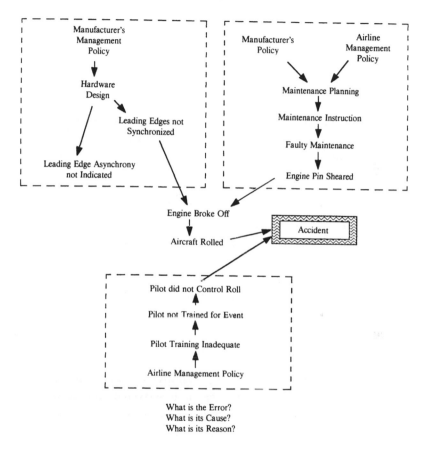

FIG. 4. Why did the accident occur?

board, the relative positions of keys on the keyboard might be the cause. If one were a psychotherapist, key positions would be irrelevant, since one would inclined to say that the typist's unconscious motivation to make an error would have manifested itself as some keyboard error regardless of the layout of the keyboard. If one were a politician intent on labor relations reform, the cause might be seen in the boss's authoritarian approach to office management. Note that to adopt one of those causes as *the* cause does not in any sense rule out the validity of the others.

WHAT IS "CAUSE?"

To a large extent a "cause" is any explanation that is adequate and acceptable in terms of the inquiry being conducted. The purpose of the inquiry, not merely the description of the events, is what determines the "cause" of an error. It is what the investigator sees as leading inevitably (or perhaps only probabilistically) from one event to another. It should also be noted that "inevitability" is always a subjective judgement by the investigator. We need to understand causes if we are to understand how errors arise. This is important both so that errors can be prevented and also so that we can have a theory and explanation of error.

Philosophical discussions have tended to distinguish between "causes" and "reasons." For that matter, in some languages there are different grammatical forms for these two words that make it impossible to talk about them in the same way. In terms of analyzing the relation between events, errors and accidents, the "reason" for the error or accident seems to be more loosely connected with what happens than does the "cause." This "looseness" may be spatial, temporal, or logical. It was indeed suggested that "causes make errors happen; reasons are how people rationalize them later."

Let us look at a migraine attack that leads to a driving error and thus to an accident. We shall assume that the physiological event in the brain that triggers the attack is in fact a random "neural accident," a change in the behavior of a group of nerve cells brought about by what seems a random biochemical alteration. Many migraine sufferers experience the first sign of an attack as the appearance of a small blind spot in the visual field. Often this goes unnoticed for some time as it grows. If, in such an early stage, the image of an overtaking vehicle fell on the blind spot just as the driver decided to pull out, his turning of the wheel would be "an error" as viewed by a external observer. The "cause" of the error could reasonably be said to have been the *neural events*, which made the road appear empty. On the other hand, suppose the driver notices the blind spot (called a "scotoma") and recognizes that it is likely to increase in size. If the driver continues to drive, and later fails to see a vehicle because of the scotoma, then we would be inclined to say that the *decision to continue driving* was an error. In particular we would say that it was a mistake (that is, an error of intention). But it is not really credible to say that the *decision* was the cause of the accident. Rather, it was the *reason* that the scotoma caused the accident.

Paradoxically, viewing errors as having many causes can mean that it becomes easier to see how to prevent errors. If there were only one cause, it would have to be rendered ineffective if the error were to be prevented. That could prove impossible or at least very difficult. But if there are many causes, then it will suffice to break the causal chain *anywhere* in order to prevent the error.

Risk Management and the Control of Errors 9

HOW DO WE PREVENT ERRORS AND ELIMINATE THEIR EFFECTS?

It goes without saying that reducing error is desirable, and when the errors imperil human beings or the environment, it is absolutely crucial. After disastrous accidents like Three Mile Island, the general public suddenly and understandably became very interested in the various aspects of human error. We as human beings *need* to know more about the things that threaten us. Yet it does not follow that reducing error will be easy, despite the increasing amount of research in this field.

If there is a random aspect, a "quantal uncertainty" about errors (as we have earlier suggested), then they can never be totally prevented. Their consequences, however, can be reduced by appropriate design, planning, training, and improvement in procedures. Certainly human factors research, as well as other branches of psychology and engineering, should be involved to that end.

PROBLEMS WITH ERROR
REDUCTION AND PREVENTION

We can, in fact, reduce both the frequency and severity of errors, although it is probably impossible that they can be eliminated altogether. In fact, when we look very closely at the relationship between erroneous and correct behavior, there a number of reasons why reducing error to a frequency and severity approaching zero may not be entirely desirable.

It is the undesirable consequences of error, not error in itself, that we must try to reduce. This may not necessarily involve a reduction of error rate. This conclusion stems from much of our earlier discussion, especially the discussion of taxonomies. The observable error behavior is not always the important error. If you pressed the wrong button, it may not be due to an erroneous movement or a flaw in the control panel design, but because you diagnosed the situation incorrectly, or did not understand an instruction. If we concentrate on reducing or eliminating *observed error*, we may wind up spending our time, resources, and money on an inappropriate part of the system, or an inappropriate change in design. After Three Mile Island there was an overemphasis on control room evaluation and redesign, but reevaluation and redesign by itself is no cure-all. The feeling that "now that the control room has been redesigned and all the recommendations implemented, the system is safe from error," is a very dangerous illusion. It is too easy and obvious to concentrate on preventing the errors that we are already familiar with, the *errors that have already occurred*. What we must do is imagine the errors that *could* happen, and think about how they can best be prevented.

There is another problem with trying to make a completely safe and errorless world, a world in which the consequences of error are always absorbed or deflected. The less often errors occur the less likely we are to expect them and the more we believe that they cannot happen. Therefore we are less prepared to cope with them, correct them, and prevent their undesirable effects when they *do* occur. It is a paradox: the more errors we make the better we can deal with them! If we are seldom called upon to intervene or exercise control, maintaining our skill levels is much more difficult. This is a well-known problem in the operation of sophisticated and highly reliable systems.

It is natural to believe that if a system is redesigned to eliminate error-inducing features, then the number and frequency of errors will decrease. That should indeed be the case, but only with respect to the *old errors*. A redesigned system is *not* the old system minus the defective features. It is a *new system*, with new opportunities for new and different errors to occur. Sometimes these will be errors that have never before occurred in the system. In a time of rapid technological change there may be errors that have never occurred in *any* system. For example, a traditional "hard wired" control room with analog displays and hundreds of manual controls provides many opportunities for certain types of error, perhaps due to confusion among controls, or because related information is conveyed by instruments that are physically too far apart. Such errors will certainly decrease if the control room is retro-fitted with computer displays and controls. New errors, however, will unquestionably arise. These might be due to the nature of electronic displays, to key punching problems, or to difficulties with the serial nature of information access. Whether or not there will be a net gain in safety and efficiency is not a simple matter to decide.

Similar problems may arise if we attempt to reduce errors by the use of check-lists, automated or printed. The effectiveness of these depends on such variables as the length of the check-list, the intricacies of procedures, the need for procedures to choose procedures, etc. Sometimes this can lead to a system so complex that operator behavior becomes slow, motivation fails, and the situation is as bad as ever.

Alarms and warnings can alert the operator to critical situations, but they are difficult to implement effectively. If only one or two systems need alarms, there is little problem; a combination of well-designed auditory and visual signals should reduce error rates. As systems get more and more complex, however, the design of alarms becomes correspondingly difficult. At Three Mile Island, *hundreds* of alarms went off within a few seconds! With so many signals going at the same time it was totally impossible to diagnose the situation in detail. It is well-known that the many auditory alarms present in commercial airplane cockpits are so distracting that pilots sometimes just turn them off. Attempts to devise "intelligent" alarms having some kind of logical priority or diagnostic ability have been underway for many years, but these are not perfected yet. *Alarms can be beneficial, but they can also overload the operator.*

The response to human error is often to fit more

interlocks, more preventive hardware, and defensive software that will prohibit certain commands. At first this seems entirely logical, but sometimes a normally undesirable tactic may be the only thing that can resolve a problem. If a pilot accidentally lowered the undercarriage of an aircraft at high speed, the landing gear could be torn off. It is easy to imagine the passage of safety legislation requiring a mechanical interlock that would prevent the landing gear being lowered except during the landing sequence. It is easy to think of such legislation, and of such a device, as good and useful. Consider, however, the following case: a few years ago a 727 hit clear air turbulence and dropped thousands of feet. The pilot, operating contrary to standard procedure, lowered the landing gear. In this situation, the undercarriage acted as an air brake and allowed him to regain control.

THE VALUE OF ERROR

There is some positive value in error. Eliminating the opportunity for error severely limits the range of possible behavior. Error can be creative, and there is a role for creative error. Trial and error learning behavior is open-ended, unlike the training methods advocated by Skinner, in which the trainer shapes the correct behavior in such a way that few if any errors occur. Indeed, a basic problem with the Skinnerian approach is that it requires the trainer to know the "correct" action in advance. In trial and error, new ways to perform tasks will be discovered. Of course, some of the time they will be unsatisfactory, slow, or inefficient. On other occasions, however, more efficient short-cuts or completely different ways of approaching the task will be found. Indeed, apart from analytic logical deduction, "trial"—with its inevitable accompaniment of error—is the *only* way that inventive approaches to a task can occur.

> *The consequences of error play a crucial role in learning and the guidance of behavior.*

It is possible that encouraging errors could benefit learning. This probably is not true for simple actions, like pressing buttons or throwing switches. When teaching diagnostic skills, or the dynamics of a complex system, however, erroneous actions throw the system into new and unexpected states. These can convey much new information, most of which is unlikely to be revealed by systematic training in "the correct way" to operate the system.

ERROR-FORGIVING DESIGN

If we are to tolerate (or even encourage) erroneous behavior, a concomitant requirement is that *the operator's environment must be tolerant. Catastrophic effects must not follow with great rapidity from an error.* The system must be error-absorbing or "forgiving" so that there will be time for the operator to benefit from the information that an error has been made. Ideally, the consequences of error should be reversible, making it possible to retrace one's steps to an earlier state. This places considerable (and unusual) demands on system design. If we consider a much broader spectrum of situations than those involving human-machine systems, it may be impossible. "Tout comprendre, c'est tout pardonner" is a maxim that can seldom be adopted in respect of errors made in human-machine interactions.

IS ERROR PREVENTION POSSIBLE?

Having drawn attention to certain unappreciated advantages of error, it remains true that in most situations of daily life many errors are undesirable. We return to the question, how can the prevalence of error be reduced?

If the occurrence of error is in some sense random, it will be difficult or impossible to prevent *a particular error* on *a particular occasion*. At the same time, many factors are known to alter the frequency of errors. While it may be impossible to prevent a specific error, we can certainly reduce the number of such errors within a given time period. Indeed, there is a rich and extensive set of choices of how to tackle these problems. They can be subsumed under the general principle: *Design should avoid error-inducing situations.* The following material discusses some of the specific ways of accomplishing this goal.

SYSTEM RELIABILITY AND REDUNDANCY

The reliability of inanimate systems can almost always be improved by redundancy and parallelism. If a subsystem has a probability P of failing during operation, then two such subsystems operating independently on the same input in parallel have a probability of simultaneous failure of P^2, three of P^3, and so on. An optical scanner may fail to detect a fault once in 1000 (10^3) times. If three such scanners operate independently, the chance of all of them failing is only 1 in a billion ($1000 \times 1000 \times 1000$, or 10^9). Any desired level of reliability can be achieved if the user is prepared to pay for redundancy.

SOCIAL INTERACTION AND ERROR

Redundancy does not always work to reduce human error. A problem that has been largely ignored in the study of human error is that behavior is seldom solitary. Many, perhaps most, of the systems with which human beings interact are composed of a number of partly autonomous subsystems. Most discussions of error concentrate on the individual, but social interactions change the equation considerably, in ways that are difficult to predict.

Errors may not decrease if an operator's task is duplicated or checked by another operator. They may just as easily increase or remain constant. For example, the operator may well make fewer errors if the second person is a quality control inspector whose comments might result in loss of pay or unemployment. On the other hand, if both job and pay are secure, the knowledge that someone else will be catching errors may lead the operator to be more lax—and commit more errors. A combination of both effects (or a genuine ability to ignore a co-worker's existence) may lead to no change. Furthermore, subtle social effects may supervene. A supervisor's instruction may be interpreted in a variety of ways.

Communication and Error

Research is also needed on the effects of communication. Probably a majority of tasks take place in a social setting, involve more than one person, and most involve as well a hierarchy of communication. This includes: person-machinecommunication (in both directions); person-to-person communication among peers; and person-to-person communication from worker to supervisor to management (and vice versa).

It is obvious that ambiguities in the communication chain from management to worker can give rise to "errors," although here again the problem of definition becomes crucial. If management demands rapid rather than accurate work, if your supervisor encourages "cutting corners" in regard to safety, are you making an error when you obey? *Yes*, from an outside, "objective" viewpoint. *No*, because you want to follow your supervisor's instructions and the policies of management. *Managerial styles and the social dynamics of the work environment are potent factors for decreasing or increasing error.*

To an operator in an inspection task, an instruction such as "try harder—make fewer errors" may have very different effects depending on the nature of "errors." Is an "error" missing a faulty component, or rejecting a sound one—a false alarm? There are a number of ambiguous areas that can affect the operator's reliability. What is the perceived "payoff" associated with misses or false alarms? What is the probability of faulty components? Both of these will necessarily affect judgement. Similarly, if management's policy is interpreted by workers to place emphasis on speed and productivity rather than accuracy and safety, the operator will change the speed-accuracy tradeoff according to this perception. A social habit or pressure from a peer group may also change such criteria. Comments from colleagues suggesting that one diagnosis of a situation is the most probable may prevent other hypotheses from being examined. Finally, communication between workers by speech, gesture, or electronic transmission devices can itself generate extra errors in the content of its message and its interpretation. *The use of multiple humans in the way multiple inaccurate components are used is not a reliable way to enhance human-machine system reliability.*

In the next few decades we shall have to contend with a whole new area of social dynamics, namely the interactions between humans and "intelligent machines," "expert systems," and similar technological innovations. Such systems may well reduce human error by absorbing some of the load inherent in knowledge and information processing. However, if their introduction results in operators' having less practice or becoming more dependent on these artificial aids, new kinds of opportunities for error and new kinds of error will appear.

Group and Individual Decisions in Error Reduction

There is a further point to be noted in light of the distinction between exogenous and endogenous errors. The probability of endogenous error may be reduced by group rather than solitary decisions, especially if the judgements can be made as independently as possible. But this will not, in general, be true of exogenous errors. A badly designed meter or a screen that reflects glare will render the data displayed just as illegible to a committee as to an individual.

There may also be differences in the value of multiple participants for various kinds of errors. Errors of substitution-action errors, and errors of judgement or intention are largely endogenous, while errors of omission are largely exogenous. This latter class will be least affected by the employment of multiple decision-makers.

ERROR THERAPY

It is crucial that we understand the origin of errors. Knowledge of the genesis of a particular error can be used to reduce its occurrence. If the distinction between "mistakes" (errors of intention) and "slips" (errors of action) can be maintained, then there may be different strategies for each type. If "mistakes" are due to misunderstanding or misinterpreting information, improved training is the remedy. If "slips" are due to an incompatibility between the desired response and the controls available, or misreading a signal because of a poor display, then a redesign of the hardware is called for.

FEEDBACK IN THE SYSTEM

Even when system design has been improved as much as possible, and training has been carried as far as possible, errors will still occur. At some point it is not economic to pursue their reduction further. Here the concept of the "forgiving system" again becomes important: the system should be designed to absorb errors. Yet, paradoxically, it is equally important that errors when made should be obvious. The actor requires feedback and knowledge of results about his or her performance, so that once an error occurs it can be corrected before its effects become serious. Feedback is essential to the control of a system that is subject to disturbances of any kind. This is particularly important when the disturbance may be caused by the operator.

For certain types of error, feedback may be extremely difficult to provide until the effects of error have propagated extensively throughout the system. Consider for example "cognitive lockup" or "cognitive tunnel vision," the tendency for an actor to select a particular hypothesis and then stick to it without looking for alternatives. How can the adoption of an incorrect hypothesis be identified and reported to the decision-maker? If it were possible for an observer or the system itself to realize that the operator is thinking incorrectly, then the operator would be unnecessary. Perhaps the most we can hope for is a system wherein there will be every opportunity and encouragement for a decision maker to think flexibly and adopt different hypotheses easily.

THE OBSERVATION OF ERRORS

Errors when perceived may generate new behavior to correct their effects. The crucial problem is to ensure that *errors are perceived when they occur.* The observability of errors may be dangerously reduced in systems with complex dynamics and long response times. Complex human-machine systems should be designed to maximize the observability of errors and minimize their effects. This may require designing systems to provide feedback of different kinds and at different levels:

- Has the action been done correctly?
- Has the expected effect occurred?
- Has the distance to the goal been decreased?

These are three levels of feedback. It is not inconceivable that an "error" in the first level might be acceptable if the third criterion is satisfied. One problem is that the automatization of behavior that comes with increasing skill reduces the role of consciousness in control, and hence reduces the use of high level goal-directed judgement. This, in turn, reduces the use of feedback at rule- and knowledge-based levels of behavior.

TRAINING AND ERROR REDUCTION

Obviously, training plays a key role in the reduction of error. Well-trained operators whose knowledge and skill are appropriate to the task will make fewer errors than unskilled operators. (We are not concerned here with philosophical subtleties of the sort raised elsewhere in this book. While it may be true in a sense that a person with limited knowledge makes fewer "errors" because he or she knows fewer correct things to do and hence can err in fewer ways, we are here talking of "erring" in the sense of *taking or omitting actions that imperil the system and its user.*)

Psychological learning theory established long ago that appropriate training reduces error, and increases the rapidity with which the learning curve approaches a performance level where further training is not necessary. When it is possible to specify the *exact* behavior required in response to *exact* information, human behavior can be shaped quite effectively. The problem is that in daily life the required behavior is often not so easily specified. Since there are frequently many ways to carry out a task, operator behavior must be goal-directed, rather than behavior-bound. In the case of complex systems *the* correct action in a given situation may not even be known, which obviously makes exact shaping of behavior impossible. The diagnosis of system states on the basis of displayed information may be difficult. There are many situations that result in similar information on the displays, and common-cause or common-mode system failures are hard to interpret. Furthermore, there is a difference between skill acquisition and skill maintenance. Operators will be initially trained up to the required skill level, after which it is assumed that they will retain their skills. That is probably true for skills that are exercised all the time, but not for those rarely used.

If operators may be required to exercise critical skills at intervals of weeks, months, or longer, a major reduction in human error may be expected from a skill maintenance program. (A very good example is the compulsory retraining of airline pilots in simulators at intervals of a few months to ensure that their ability to deal with emergencies is maintained.)

SKILL AND KNOWLEDGE

Some kinds of errors, particularly those of physical activity, are very obvious even to the actor. An incorrect turn of the steering wheel or handlebar, an incorrect note on the piano, a missed step in dancing—all these have immediate and striking consequences. Even when the behavior is almost completely automatic, errors will often—though not always—be noted and attempts will be made to correct them. If skill is defined as goal-directed behavior we can say that it is inherently error-correcting, since consciously or unconsciously an individual's actions are continually modified so as to approach the goal even more closely. If this were not the case, the actor would not be "skilled."

The relations between skill, knowledge, and the following of rules or instructions are not immediately clear in this context; there is ample evidence that skill is not directly aided by knowledge. Instructions are more likely to be followed correctly if the reasons for the instruction are known. (Note, incidentally, that this is not the same as understanding the physical laws governing the behavior of the system.) Errors are more likely to be recorded if the reasons for their documentation are made clear. When a system is relatively simple, the operator will have a smaller burden of understanding, and errors will be reduced or noted more efficiently.

It is important to note that the relation between skill and knowledge in the context of error situations is not fully understood. Much research is still needed on this relationship. Skill is the *ability* to carry out a task. Knowledge is the possession of *information, facts, and understanding* about a task. High levels of skill tend to produce low levels of error: indeed part of the very definition of skill is that the behavior is virtually error-free. However, the acquisition of skill and the acquisition of knowledge are not identical. A person may know a lot about a task and still not be able to carry it out (hence the worry many employers have about employees who are "too academic"). A person may have very high levels of skill, especially in perceptual-motor tasks, and have little expressible knowledge; few bicyclists can explain the dynamics of the vehicle or are aware of the actions they take to ensure stability while riding.

Most training is concerned with skill development. Evidence from behavioral laboratory studies suggests that providing an operator with knowledge does little to assist the acquisition of skill. Halfway between the acquisition of unconscious automatic skills and the manipulation of conscious knowledge, we may provide rules for the operator to follow. Insofar as operating rules or procedures are clearly expressed, and based on an accurate task analysis, errors of *decision-making* should largely be abolished. The remaining errors will be in *choices of procedure*, and in *slips of action* while carrying them out.

One problem about knowledge and skill arises even when operating procedures are provided. Frequently manuals and procedures are written by the designers of the system or other experts. Such people have an enormous amount of implicit knowledge of the purpose of the system and its components, and *such knowledge is not available to the operator*. It has been demonstrated that power plant operators, in some instances, did not know the meanings of some of the words used in the operating procedures. What was the engineer who wrote the procedures trying to do? What is the operator expected to know? Why was the system configured in this way? It seems likely that an important way to reduce error is to ensure that the operator understands the intentions of the designer.

Minimizing the Impact of Error

After everything has been done to reduce errors, though, they will still occur. When they do, something must be done to minimize their impact.

> *First, the system should be forgiving; it should absorb errors at least for a time.*

> *Second, the operator should be trained to admit the possibility of error and acknowledge it.* This will free him or her to take remedial action.

Third, in the event of error some kind of feedback must be given so that it is possible for the operator to recognize (even unwillingly) that an error has occurred. This requirement is difficult to fulfill for errors of omission, for cognitive or intellectual errors, and especially for errors of intention or planning. Standardization, automation, artificial intelligence, all may help. But the latter two also distance the operator from the physical system, and make it harder for that person to realize what is happening.

Human error rates can be reduced to as low a level as desired, at some unknown cost. The occurrence of a *particular* error at some *particular* instant, however, cannot absolutely be prevented.

SUMMARY

In this chapter, our attempt to discuss methods of error-reduction seems on reflection to have ended up as a list of potential sources of error. Probably, this is a fair reflection of our state of knowledge. It is clear that in designing any task all possible basic information available from ergonomics and human factors standards should be used. Firstly, there is a great deal of classical "knobs and dials" design wisdom that will reduce error if it is adopted. Secondly, although much is known about how to write operating procedures and instructions to reduce the probability of error, this is hardly ever put to use by business, industry, or government. Thirdly, as we have seen, it is important to take into account the dynamics of management. In general an authoritarian management with a punitive attitude to error will lead to the concealment of errors, and thus others cannot benefit from reports by those who commit errors. (Contrast this with, for example, the NASA/FAA scheme for "no fault" reporting of pilot errors.)

As mentioned earlier, when people operate complex systems and trouble arises, a major problem is their tendency to "lock up" on an early hypothesis, preventing them from considering others. It would seem that systematic training in the consideration of alternatives—training in "how to change one's mind"—could reduce such persistence errors. Certainly it would be valuable for another human or a computer aid to backtrack to the beginning of a decision sequence and encourage the operator to start again with a new initial set of assumptions. It is certainly desirable that operators be given skill-maintenance training in a systematic way. *Part-task simulation* is a relatively cheap and effective way to do this. With such new technologies as computers and video discs becoming widely available, it should be increasingly cost-effective. There is also evidence that in certain skills (such as fault diagnosis) generic training is possible; there seems to be considerable transfer of skill from one task to another, so that context-free training may be useful to some extent. There is a huge literature dating back to the 1930's on transfer of training in laboratory tasks, but our understanding of the conditions in *real life* under which such transfer occurs is rudimentary.

One thing is clear. Mere exhortation to "try harder" is not an efficient way to reduce error. Short of malice, no one *wants* to make errors. Good design is essential. The extensive application of ergonomics and anthropometrics is essential. Effective instructions, procedures, and rules are essential. The application of management and social dynamics appropriate to the task and the culture is essential. Good initial training and skill maintenance are essential. Although it is largely ignored by designers, engineers, computer people, management, industry, business, and government, information on all the above is available in ample quantities from the human factors community. Those industries where this knowledge is used systematically (such as the aerospace industry) have a markedly good record of reducing the impact of errors on system performance.

A considerable body of knowledge exists. We are, however, far from applying it effectively to the full range of errors from crucial to trivial—from the operator's misdiagnosis at a nuclear power plant, to the person who pours oatmeal into the washing machine in place of detergent.

Epilogue: Further Commentaries on Error

10

"ALL THE KING'S MEN"—ORLINDO PEREIRA

Scientific analysis is always a reductionistic enterprise based on arbitrarily chosen starting points. The contributors to the Error Conference reported in this book tended to envisage the many problems discussed within the framework of a one man—one machine system.

In reality, in many working situations, including control tasks, the depiction should be of a one team—one machine system. If, for example, the human controller is not alone in the control room, his or her behavior is expected to be different from that which he or she shows when alone. This is, obviously, an old question in social psychology.

It is stated on page 30 that if an accident happens because of one person's error and another person's mistake, there is a cause for both the error and the mistake, but not for their simultaneity, which is the cause of the accident. If we take into consideration group behavior as a necessary instance of the system under analysis, that claim may not hold. It certainly will

not hold if the behavior of both individuals is interde-
pendent, as is the case in all group structures and
dynamics. The risky shift and groupthink are but two
examples taken from social psychological research in
which group decision blunders occur. The latter is
more prone to happen when the system is designed, the
former when the controller is a member of a team and
calls attention to the possible role of leadership in the
origin of both errors and mistakes.

If "the use of multiple humans in the way multiple
inaccurate components are used is not a reliable way to
enhance human-machine system reliability," (Chapter
9, p. 120) it is because interpersonal behavior is not
something you may add to a one-man, one-machine
system. If you try to do so, what really happens is a
complete system reshaping.

I would like to illustrate the above point using
research results (Pereira & Jesuino, 1988) taken from
a series of studies aimed at clarifying the relationships
between stress and leadership in Marines.

In the present volume stress is identified as a condi-
tion of the organism that is supposed to increase the
probability of error. Our studies show that it may not
be the case. Stress is not bad, per se, as training may not
be good, per se. We have found that, in demanding,
new (i.e., stressful) tasks, Marine leaders may withhold
support to the followers in order to increase their strain
(reported symptoms of stress) in order to increase their
performance level (i.e., so that they make fewer execution errors). In doing so, leaders also decrease the
satisfaction of their followers. On the contrary, if

leaders give more support to the followers, the latter's strain decreases, their satisfaction is higher—but their performance deteriorates.

Our results also show that authoritarian, or for that matter formal, leaders (i.e., leaders who strictly and rigidly enforce organizational specifications) are prone to increase strain, decrease satisfaction, and compromise the performance of their followers, thus increasing the probability of error.

What seems to make a difference in the behavior of the leaders who are able to manage a demanding performance from the followers, and that of the formal authoritarian ones, is the fact that the former use discretionary leadership practices. In other words, they are able to use power and influence that are beyond the organization's specifications, and they do that on a personal basis, discriminating each one of their subordinates and maintaining with each one a personal relationship.

Perhaps the picture I depict is too "frightening" to the systems engineer, used to exercising precise control upon the system's "parts." We may suspend judgment up to the moment in which we have specific research results, aimed at measuring errors in complex human-human-machine interfaces. That the question is relevant, I have no doubt. It remains to be seen if "all the king's men" will be able to put the system together again.

Orlindo Gouveia Pereira
The New University of Lisbon

"AN UPDATE ON HUMAN
ERROR"—JOHN WREATHALL

In my most humble opinion, the understanding of *manifestations* of human error has developed significantly since 1983, but our understanding of its *origins* has not.

In the time since 1983, we have experienced accidents of a kind that was not under active consideration by the safety community; these have had an influence on how we see human error influencing, and being influenced by, the technical setting of human beings. Up to the time of the conference, we were strongly conditioned by concerns of misunderstandings, such as the 1979 accident at Three Mile Island. Taxonomies of errors such as the classical "slip vs. mistake" are very powerful for such events.

Since then, we have had several major accidents that result from more complex error mechanisms that are apparently organizational or social in origin. Examples include Chernobyl, Kings Cross, Herald of Free Enterprise, Bhopal, and Challenger. In these cases, the human errors were organizational in nature, with no one or two individuals accountable. This has given rise to the giving of attention to organizational assessment, with concepts like "resident pathogens" (Reason), "pathological vs. calculative vs. generative organizations" (Westrum), and "programmatic performance indicators of organizations" (Wreathall et al.) coming into vogue. Of course, accidents involving organizational error have always occurred, such as Flixborough, Air Florida, and Aberfan. However, the perception of the technical community (at that time) seemed to have played down these events in favor of focusing on individuals and their errors.

In terms of the origins of errors, I continue to believe that, in practice, we must largely treat their occurrence as a random variable that can be only moderately modified by external factors such as training and task aids. The use and assessment of these techniques will be promoted by the so-called "generative" organizations. However, the underlying mechanisms of error generation are as much a mystery now as they ever were.

References

Altman, J. W. *Classification of Human Error*, in Symposium on the Reliability of Human Performance in Work, W. B. Askren, (Ed), Ann. Conv. Amer. Psych. Assn., 1966.

Bartlett, F. C., *Fatigue Following Highly Skilled Work*, Proc. Roy. Soc., v 131, 248–257, London, 1943.

Bartlett, F. C., *Remembering: A Study in Experimental and Social Psychology*, Cambridge University Press, 1932.

Chapman, L. J., & J. P. Chapman, *Genesis of Popular but Erroneous Diagnostic Observations*, J. Abn. Psy., 72, 193–194, 1967.

Fischhoff, B., *Hindsight Does Not Equal Foresight; the Effect of Knowledge on Judgment under Uncertainty*, J. Exp. Psy.; Human Perception & Performance, 1, 288–299, 1975.

Freud, S. *Psychopathology of Everyday Life*, Ernest Benn, London, 1914.

Kahneman, D., P. Slovic & A. Tversky, *Judgment under Uncertainty; Heuristics and Biases*, Erlbaum, Hillsdale, N. J.,1982.

Kollarits, J., *Beobachtungen über Dyspraxien (Fehlhandlungen)*, Arch. für die Gesamte Psychologie, v 99, 305–399, 1937.

Norman, D. A., *The Psychology of Everyday Things,* Basic Books, New York, 1988.

Pereira, O. G. & J. C. Jesuino, *Coping with Stress in Military Settings: Marines at war and peace*, in D. Canter et al. (eds), Environmental Social Psychology, NATO ASI Series, vol D-45, Klüver, 1988.

Reason, J. *Human Error.* Cambridge University Press, 1990.

Senders, J. W. & A. J. Sellen, *A Quantitative Model of Human*

Error, Proc. Amer. Stat. Ass. Annual Mtg., Monterey, CA, 1987.

Spearman, C. *The Origin of Error*. Journal of General Psychology, 1, 29–53, 1928.

Swain, A., & H. E. Guttmann, *Handbook of Human Reliability Analysis with Emphasis on Nuclear Power Plant Applications*, NUREG/CR-1278, U. S. Nuclear Regulatory Commission, Washington, 1983.

Wason, J. P. & P. Johnson-Laird, *Psychology of Reasoning; Structure and Content*, Batsford, London, 1972.

Appendix I

LIST OF PARTICIPANTS
THE CONFERENCE AT COLUMBIA
FALLS, MAINE
7–9 JULY, 1980

Prof. Ward Edwards
Social Science Research Institute
Univ. of Southern California
Los Angeles, CA, U.S.A.

Dr. David Embry
Human Reliability Associates, Ltd.
1 School House, Higher Lane
Dalton, Parbold
Lancashire, U.K.

Prof. Patrick J. Foley
Dept. of Industrial Engineering
Univ. of Toronto
Toronto, Ontario, Canada

Prof. Brian Gaines
Dept. of Computer Science
Univ. of Calgary
Calgary, Alberta, Canada

Prof. Daniel Kahneman
Dept. of Psychology
Univ. of California at Berkeley
Berkeley, CA, U.S.A.

Dr. Marcel Kinsbourne
Dept. of Behavioral Neurology
Shriver Center
200 Trapelo Road
Waltham, MA, U.S.A.

Prof. John Lyman
MMEE Laboratory
Univ. of California at Los Angeles
Los Angeles, CA, U.S.A.

Prof. Neville P. Moray
Dept. of Mech. & Ind. Eng.
Univ. of Illinois
Urbana, IL, U.S.A.

Prof. Donald A. Norman
Center for Cognitive Studies
Univ. of California at San Diego
La Jolla, CA, U.S.A.

Prof. James Reason
Dept. of Psychology
Univ. of Manchester
Manchester, U.K.

Prof. Frank Restle (dec.)
Dept. of Psychology
Indiana Univ.
Bloomington, IN, U.S.A.

Prof. John W. Senders
Dept. of Mech. Engineering
Univ. of Maine
Orono, ME, U.S.A.

Prof. Thomas B. Sheridan
Dept. of Mech. Engineering
Mass. Inst. of Technology
Cambridge, MA, U.S.A.

Prof. Lawrence Stark
Dept. of Eng. Sciences
Univ. of California at Berkeley
Berkeley, CA, U.S.A.

Dr. Alan D. Swain
Sandia Corp.
712 Sundown Place, S.E.
Albuquerque, NM, U.S.A.

Prof. Anne M. Treisman
Dept. of Psychology
Univ. of California at Berkeley
Berkeley, CA, U.S.A.

Dr. William Verplank
ID TWO
1527 Stockton Street
San Francisco, CA, U.S.A.

Prof. Boardman C. Wang
Dept. of Anaesthesiology
School of Medicine
New York Univ.
New York, NY, U.S.A.

Prof. Laurence Young
Dept. of Aeronautics & Astronautics
Mass. Inst. of Technology
Cambridge, MA, U.S.A.

Appendix II

Dr. David Embry
Human Reliability Associates, Ltd.
1 School House, Higher Lane
Dalton, Parbold
Lancashire, U.K.

Dr. Klaus-Martin Goeters
DFVLR-Av. Psych.
D-2000, Hamburg 63
Fed. Rep. of Germany

Dr. Martine Griffon-Fouco
Dept. de Expl. Sûreté Nucleaire
Electricité de France
3, rue de Messine
75008 Paris, France

Dr. Erik Hollnagel
OECD Halden Reactor Project
Inst. for Energiteknik
Postbox 173-N-1751
Halden, Norway

Prof. Ezra S. Krendel
Dept. of Statistics
Wharton School
Univ. of Pennsylvania
Philadelphia, PA, U.S.A.

Prof. Elizabeth F. Loftus
Dept. of Psychology
Univ. of Washington
Seattle, WA, U.S.A.

Dr. Giovanni Mancini
JRC-Ispra
21020 Ispra, Italy

Mr. Duane T. McRuer
Systems Technology, Inc.
13766 S. Hawthorne Blvd.
Hawthorne, CA, U.S.A.

Prof. Neville P. Moray
Dept. of Mech. & Ind. Eng.
Univ. of Illinois
Urbana, IL, U.S.A.

Prof. Donald A. Norman
Center for Cognitive Studies
Univ. of California at San Diego
La Jolla, CA, U.S.A.

Prof. Orlindo Gouveia-Pereira
Universidade Nova de Lisboa
Faculdade de Economia
Rua Marquès da Fronteira, 20
1000 Lisboa, Portugal

Prof. Jens Rasmussen
RIS0 National Laboratory
DK-4000
Roskilde, Denmark

Prof. James Reason
Dept. of Psychology
Univ. of Manchester
Manchester, U.K.

Dr. William B. Rouse
Search Technology, Inc.
4725 Peachtree Corners Circle, #200
Norcross, GA, U.S.A.

Prof. William Ruddick
Dept. of Philosophy
New York Univ.
New York, NY, U.S.A.

Prof. John W. Senders
Dept. of Mech. Engineering
Univ. of Maine
Orono, ME, U.S.A.

Prof. Thomas B. Sheridan
Dept. of Mech. Engineering
Mass. Inst. of Technology
Cambridge, MA, U.S.A.

Dr. Alan D. Swain
Sandia Corp.
712 Sundown Place, S.E.
Albuquerque, NM, U.S.A.

Dr. David Taylor
Dept. of Psychology
Southampton Univ.
Southampton, U.K.

Prof. Willem A. Wagenaar
Dept. of Psychology
Univ. of Leiden
2500 RA Leiden
The Netherlands

Prof. David Woods
Dept. of Ind. & Systems Eng.
The Ohio State Univ.
Columbus, OH, U.S.A.

Dr. John Wreathall
Systems Analysis Internat'l Corp.
2941 Kenny Road, #210
Columbus, OH, U.S.A.

Index

147

Printed and bound by CPI Group (UK) Ltd, Croydon, CR0 4YY

23/10/2024

01778237-0002